I0495725

Light-Duty Automotive Technology, Carbon Dioxide Emissions, and Fuel Economy Trends: 1975 Through 2009

Compliance and Innovative Strategies Division
and
Transportation and Climate Division

Office of Transportation and Air Quality
U.S. Environmental Protection Agency

NOTICE

This technical report does not necessarily represent final EPA decisions or positions. It is intended to present technical analysis of issues using data that are currently available. The purpose in the release of such reports is to facilitate the exchange of technical information and to inform the public of technical developments.

United States
Environmental Protection
Agency

EPA420-R-09-014
November 2009

Table of Contents

Page Number

I.	Executive Summary	i
II.	Introduction	1
III.	Fuel Economy Trends	5
IV.	Carbon Dioxide Emissions Trends	17
V.	Fuel Economy Trends by Vehicle Type, Size, and Weight	30
VI.	Fuel Economy Technology Trends	49
VII.	Marketing Groups and Fuel Economy	76
VIII.	Characteristics of Fleets Comprised of Existing Fuel-Efficient Vehicles	86
IX.	References	95

Table of Contents, continued

Appendices

APPENDIX A -- Database Details and Calculation Methods

APPENDIX B -- Model Year 2009 Nameplate Fuel Economy Listings

APPENDIX C -- Fuel Economy Distribution Data

APPENDIX D -- Fuel Economy Data Stratified by Vehicle Type

APPENDIX E -- Fuel Economy Data Stratified by Vehicle Type and Size

APPENDIX F -- Car Fuel Economy Data Stratified by EPA Car Class

APPENDIX G -- Fuel Economy Data Stratified by Vehicle Type and Weight Class

APPENDIX H -- Fuel Economy Data Stratified by Vehicle Type and Drive Type

APPENDIX I -- Fuel Economy Data Stratified by Vehicle Type and Transmission Type

APPENDIX J -- Fuel Economy Data Stratified by Vehicle Type and Cylinder Count

APPENDIX K -- Fuel Economy Data Stratified by Vehicle Type, Engine Type and Valves Per Cylinder

APPENDIX L -- Fuel Economy Data Stratified by Vehicle Type and Marketing Group

APPENDIX M -- Fuel Economy by Marketing Group, Vehicle Type and Weight Class

APPENDIX N -- Fuel Economy and Ton-MPG by Marketing Group, Vehicle Type and Size

APPENDIX O -- MY2009 Fuel Economy by Vehicle Type, Weight and Marketing Group

APPENDIX P -- Fuel Economy Data Stratified by Marketing Group and Vehicle Type

APPENDIX Q -- Characteristics of Fleets Comprised of Fuel Efficient Vehicles

I. Executive Summary

Introduction

This report summarizes key trends in carbon dioxide (CO_2) emissions, fuel economy and technology usage related to model year (MY) 1975 through 2009 light-duty vehicles sold in the United States. Light-duty vehicles are those vehicles that EPA classifies as cars or light-duty trucks (sport utility vehicles or SUVs, vans, and pickup trucks with gross vehicle weight ratings up to 8500 pounds). The data in this report supersede the data in previous reports in this series.

On September 15, 2009, EPA proposed the first-ever light-duty vehicle greenhouse gas emissions standards, under the Clean Air Act, for MY2012-2016 (74 Federal Register 49454, September 28, 2009). These proposed standards are part of a new, harmonized National Policy that also includes proposed corporate average fuel economy (CAFE) standards for the same years by the Department of Transportation's National Highway Traffic Safety Administration (NHTSA). Accordingly, while past reports in this series focused exclusively on fuel economy data, this year's report provides some key industry-wide tailpipe CO_2 emissions data for the 1975 – 2009 time series as well. Tailpipe CO_2 emissions data represent 90 to 95 percent of total light-duty vehicle greenhouse gas emissions. Section IV of this report discusses the CO_2 emissions data in more detail and also provides guidance for how readers can calculate CO_2 emissions values, not shown in Section IV, that are equivalent to other fuel economy values in this report.

Since 1975, overall new light-duty vehicle CO_2 emissions have moved through four phases:

1. A rapid decrease from 1975 through 1981;
2. A slower decrease until reaching a valley in 1987;
3. A gradual increase until 2004; and
4. A decrease for the five years beginning in 2005.

The projected fleetwide average real world MY2009 light-duty vehicle CO_2 emissions level is 422 grams per mile (g/mi). The fleetwide average MY2008 value is 424 g/mi. The MY2008 value is essentially a final value as the database for 2008 includes formal production data for nearly the entire MY2008 fleet, while the projected MY2009 value is based on pre-model year production projections provided by automakers and are therefore much more uncertain. Actual MY2009 sales are expected to be 30 to 40 percent lower than the projected MY2009 production volumes provided by automakers to EPA in the spring and summer of 2008. At this time, it is not possible to predict whether the market turmoil in 2009 will yield an actual CO2 emission value that is higher or lower than the preliminary MY2009 value reported here. The preliminary 422 g/mi value for model year 2009 represents a 39 g/mi, or eight percent, decrease relative to the 461 g/mi value for 2004, which was the highest CO_2 emissions value since 1980.

Since fuel economy has an inverse relationship to tailpipe CO_2 emissions, overall new light-duty vehicle fuel economy has moved through four "opposing" phases:

1. A rapid increase from 1975 through 1981;
2. A slower increase until reaching its peak in 1987;
3. A gradual decline until 2004; and
4. An increase for the five years beginning in 2005.

The projected fleetwide average real world MY2009 light-duty vehicle fuel economy is 21.1 miles per gallon (mpg), while the fleetwide average MY2008 value is 21.0 mpg. Again, EPA has much greater confidence in the MY2008 value, which is 0.2 mpg higher than the value that we projected for MY2008 in last year's report based on pre-model year production volume projections. The fact that the revised MY2008 value is higher than the preliminary value in last year's report is to be expected given that gasoline prices peaked in spring and summer of 2008. There is much less certainty associated with the projected MY2009 value of 21.1 mpg as it is based on pre-model year production projections provided by automakers, and 2009 has continued to be a year of turmoil in the automotive market. It is impossible to predict whether actual MY2009 fuel economy will be higher or lower than the preliminary MY2009 value. The projected model year 2009 value of 21.1 mpg represents a 1.8 mpg, or nine percent, increase over the 19.3 mpg value for 2004, which was the lowest fuel economy value since 1980.

The CO_2 emissions and fuel economy values in this report are either *adjusted* (ADJ) EPA "real-world" estimates (provided to consumers), or unadjusted EPA *laboratory* (LAB) values. All CO_2 emissions and fuel economy values in this report are adjusted values unless explicitly identified as laboratory data. All combinations of adjusted or laboratory, and CO_2 emissions or fuel economy values, may be reported as city, highway, or, most commonly, as *composite* (combined city/highway, or COMP). In 2006, EPA revised the methodology by which EPA estimates adjusted fuel economy to better reflect changes in driving habits and other factors that affect fuel economy such as higher highway speeds, more aggressive driving, and greater use of air conditioning. This is the third report in this series to reflect this new real-world fuel economy methodology, and every adjusted fuel economy value in this report for 1986 and later model years is lower than values in pre-2007 reports in this series. To reflect that these changes did not occur overnight, these new downward adjustments are phased in, gradually, beginning in 1986, and for 2005 and later model years the new adjusted composite fuel economy values are, on average, about six percent lower than under the methodology used by EPA in older reports. This same methodology is used to generate adjusted CO_2 emissions values as well. See Appendix A for more details.

Because the underlying methodology for generating unadjusted laboratory CO_2 emissions and fuel economy values has not changed since this series began in the mid-1970s, they provide an excellent basis for comparing long-term CO_2 and fuel economy trends from the perspective of vehicle design, apart from the factors that affect real-world driving that are reflected in the adjusted values. Laboratory composite values represent a harmonic average of 55 percent city and 45 percent highway operation, or "55/45." For 2005 and later model years, unadjusted laboratory composite CO_2 emissions values are, on average, about 20 percent lower than adjusted composite CO_2 values, and unadjusted laboratory composite fuel economy values are, on average, about 25 percent greater than adjusted composite fuel economy values. The projected MY2009 unadjusted laboratory composite values of 337 g/mi and 26.4 mpg represent a record low for CO_2 emissions and an all-time high for fuel economy.

While EPA establishes vehicle CO_2 emissions standards, NHTSA has the overall responsibility for the CAFE program. For 2009, the CAFE standards are 27.5 mpg for cars and 23.1 mpg for light trucks (for light trucks, individual manufacturers can choose between the fixed, unreformed 23.1 mpg standard and a reformed vehicle footprint-based standard which yields different compliance levels for each manufacturer). In March 2009, NHTSA promulgated new footprint-based CAFE standards for MY2011, for which NHTSA projected average industry-wide compliance levels of 30.2 mpg for cars (including a 27.8 mpg alternative minimum standard for domestic cars for all manufacturers) and 24.1 mpg for light trucks. EPA provides laboratory composite fuel economy data, along with alternative fuel vehicle credits and test procedure adjustments, to NHTSA for CAFE enforcement. Because of real world adjustments, alternative fuel vehicle credits, and test procedure adjustments, current NHTSA CAFE values are a minimum of 25 percent higher than EPA adjusted fuel economy values.

Characteristics of Light Duty Vehicles for Four Model Years

	1975	1987	1998	2009
Adjusted CO_2 Emissions (g/mi)	679	405	443	422
Adjusted Fuel Economy (mpg)	13.1	22.0	20.1	21.1
Weight (lbs.)	4060	3220	3744	4108
Horsepower	137	118	171	225
0 to 60 Time (sec.)	14.1	13.1	10.9	9.5
Percent Truck Sales	19%	28%	45%	49%
Percent Front-Wheel Drive	5%	58%	56%	55%
Percent Four-Wheel Drive	3%	10%	20%	27%
Percent Multi-Valve Engine	-	-	40%	79%
Percent Variable Valve Timing	-	-	-	65%
Percent Cylinder Deactivation	-	-	-	9%
Gasoline-Direct Injection	-	-	-	3.5%
Percent Turbocharger	-	-	1.4%	3.1%
Percent Manual Trans	23%	29%	13%	6%
Percent Continuously Variable Trans	-	-	-	8%
Percent Hybrid	-	-	-	1.8%
Percent Diesel	0.2%	0.2%	0.1%	0.5%

November 2009

Highlight #1: Carbon Dioxide Emissions Decreases and Fuel Economy Increases Over the Last 5 Years Reverse the Long-Term Trend From 1987 through 2004.

Average adjusted composite CO_2 emissions have decreased from 461 g/mi in MY2004 to a projected level of 422 g/mi in MY2009, accounting for a 39 g/mi and 8 percent decrease. The preliminary MY2009 adjusted composite fuel economy value of 21.1 mpg represents a 1.8 mpg, or 9 percent, increase over MY2004. Actual MY2009 values will likely differ from these preliminary MY2009 values, but it is impossible to know the direction or magnitude of any changes. For both CO_2 emissions and fuel economy, the last 5 years reverse a longer-term trend over the period 1987 through 2004 and essentially return CO_2 emissions and fuel economy levels to those of the early 1980s.

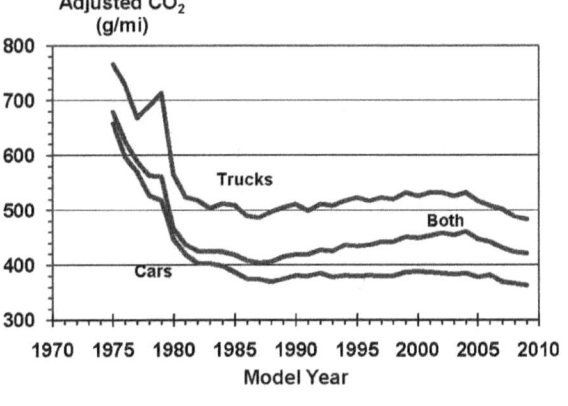

MY2009 unadjusted laboratory composite values, which reflect vehicle design considerations only and do not account for the many factors which affect real world CO_2 emissions and fuel economy performance, are at an all-time low for CO_2 emissions (337 g/mi) and a record high for fuel economy (26.4 mpg).

iv   November 2009

Highlight #2: Trucks Continue To Represent About Half of New Vehicle Production.

Light trucks, which include SUVs, vans, and pickup trucks, have accounted for about 50 percent of the U.S. light-duty vehicle market since MY2002. After two decades of constant growth, light truck market share has been relatively stable from 2002 through 2009. The MY2009 light truck market share is projected to be 49 percent, based on pre-model year production projections by automakers.

Historically, growth in the light truck market was primarily driven by the explosive increase in the market share of SUVs (EPA does not have a separate category for crossover vehicles and classifies many crossover vehicles as SUVs). The SUV market share increased from six percent of the overall new light-duty vehicle market in MY1990 to about 30 percent of vehicles built each year since 2004. By comparison, market shares for both vans and pickup trucks have declined since 1990, with van market share falling by about one-half from 10 percent to five percent. The increased overall market share of light trucks, which in recent years have averaged 120 – 140 g/mi higher CO_2 emissions and 6 – 7 mpg lower than cars, accounted for much of the increase in CO_2 emissions and decline in fuel economy of the overall new light-duty vehicle fleet from MY1987 through MY2004.

Sales Fraction by Vehicle Type
(Annual Data)

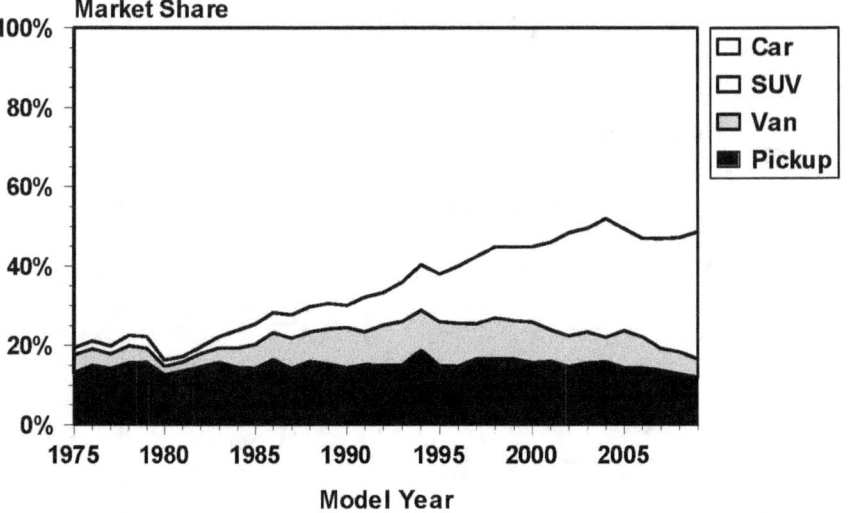

Highlight #3: Technological Innovation Since 2005 Has Resulted in Lower CO_2 Emissions, Higher Fuel Economy and Greater Performance.

> *Automotive engineers are constantly developing more advanced and efficient vehicle technologies. From 1987 through 2004, on a fleetwide basis, this technology innovation was utilized exclusively to support market-driven attributes other than CO_2 emissions and fuel economy, such as vehicle weight (which supports vehicle content and features), performance, and utility. Beginning in MY2005, technology has been used to increase both fuel economy (which has reduced CO_2 emissions) and performance, while keeping vehicle weight relatively constant.*

Vehicle weight and performance are two of the most important engineering parameters that help determine a vehicle's CO_2 emissions and fuel economy. All other factors being equal, higher vehicle weight (which supports new options and features) and faster acceleration performance (e.g., lower 0-to-60 mile-per-hour acceleration time), both increase a vehicle's CO_2 emissions and decrease fuel economy. Average vehicle weight and performance had increased steadily from the mid-1980s through 2004.

Average light-duty vehicle weight has been fairly constant since 2004, with a small increase in weight of cars offset by a small decrease in truck market share. Average fleetwide performance has continued to improve just about every year. The projection for MY2009 is for an increase in both vehicle performance and weight.

Weight and Performance
(Annual Data)

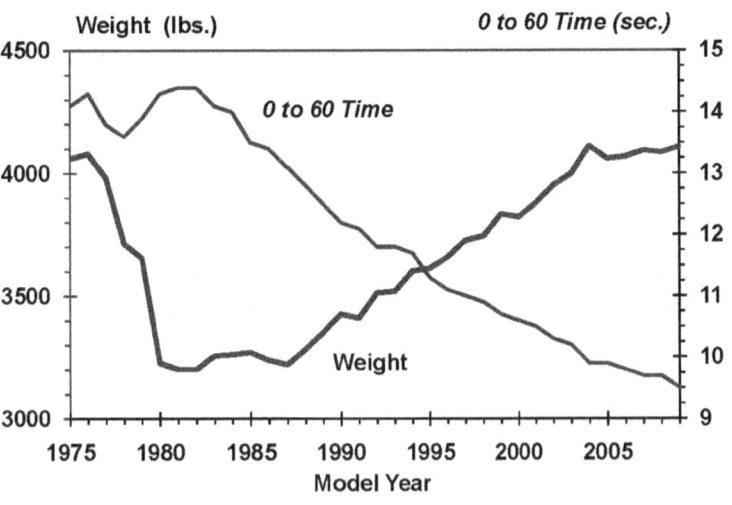

Highlight #4: Many Marketing Groups Are Increasing Fleetwide Fuel Economy, Resulting in Lower CO_2 Emissions.

> *Seven of the nine highest-selling marketing groups increased fuel economy (which also reduced CO_2 emissions) from MY2007 to MY2008, the last two years for which we have solid information based on final CAFE reports. Preliminary values suggest that four of the nine marketing groups will increase fuel economy (thereby reducing CO_2 emissions) in MY2009, and one marketing group will maintain constant levels, based on projected production provided to EPA by automakers prior to the start of the model year. Actual MY2009 values will likely be different than the preliminary MY2009 values reported here.*

In MY2008, the last year for which EPA has essentially complete formal production data, Honda had the lowest fleetwide adjusted composite CO_2 emissions (and highest fuel economy) performance, followed closely by Hyundai-Kia. Chrysler had the highest CO_2 emissions (and lowest fuel economy), with Ford having slightly lower CO_2 emissions. Chrysler had the biggest absolute improvement from MY2007 to MY2008, with an 19 g/mi, or 4.0 percent, reduction in fleetwide CO_2 emissions, followed by Hyundai-Kia with a 14 g/mi and 3.6 percent reduction in CO_2 emissions.

Preliminary MY2009 values suggest that Honda will continue to have the lowest fleetwide CO_2 emissions (and highest fuel economy), followed closely by Hyundai-Kia and Toyota. Chrysler is projected to have the highest MY2009 CO_2 emissions, reversing most of its gains from the previous year. Ford is projected to show the largest CO_2 reductions, with its projected MY2009 CO_2 emissions being 37 g/mi lower than MY2007 and 25 g/mi lower than MY2008. Ford and General Motors are the two marketing groups that showed improvement in MY2008 and are projected to do so again in MY2009.

MY2007 – 2009 Marketing Group Fuel Economy and Carbon Dioxide Emissions
(Adjusted Composite Values)

Marketing Group	<-------- MY2007 -------->		<-------- MY2008 -------->		<-------- MY2009 -------->	
	Fuel Economy (mpg)	CO_2 (g/mi)	Fuel Economy (mpg)	CO_2 (g/mi)	Fuel Economy (mpg)	CO_2 (g/mi)
Honda	23.3	382	23.9	372	23.6	376
Hyundai-Kia	22.9	388	23.7	374	23.4	380
Toyota	23.3	382	22.8	389	23.2	383
Volkswagen	21.9	405	22.3	398	22.8	398
Nissan	21.3	418	21.9	406	21.6	411
BMW	21.5	415	21.2	419	21.6	412
General Motors	19.2	463	19.7	452	19.9	447
Ford	18.9	471	19.4	459	20.5	434
Chrysler	18.6	479	19.3	460	18.7	476
All	**20.6**	**432**	**21.0**	**424**	**21.1**	**422**

Important Notes with Respect to the Data Presented in This Report

Most of the CO_2 emissions and fuel economy values in this report are a single *adjusted* composite (combined city/highway) CO_2 emissions or fuel economy value, consistent with the real-world estimates for city and highway fuel economy provided to consumers on new vehicle labels, in the EPA/DOE *Fuel Economy Guide*, and in EPA's *Green Vehicle Guide*.

This 2009 report supersedes all previous reports in this series, which date back to the early 1970s. In general, users of this report should rely exclusively on data in this 2009 report, which covers the years 1975 through 2009, and not try to make comparisons to data in previous reports in this series. There are at least two reasons for this.

One, EPA revised the methodology for estimating real-world fuel economy values in December 2006. This is the third report in this series to reflect this new real-world fuel economy methodology, and every adjusted (ADJ) fuel economy value in this report for 1986 and later model years is lower than given in reports in this series prior to the 2007 report. Accordingly, adjusted fuel economy values for 1986 and later model years should not be compared with the corresponding values from pre-2007 reports. These new downward adjustments are phased in, linearly, beginning in 1986, and for 2005 and later model years the new adjusted composite (combined city/highway) values are, on average, about six percent lower than under the methodology previously used by EPA. See Appendix A for more in-depth discussion of this new methodology and how it affects both the adjusted fuel economy values for individual models and the historical fuel economy trends database. This same methodology is used to calculate adjusted CO_2 emissions values as well.

Two, when EPA changes a marketing group definition to reflect a change in the industry's current financial arrangements, EPA makes the same adjustment in marketing group composition in the historical database as well. This maintains a consistent marketing group definition over time, which allows the identification of trends over time. On the other hand, it means that the database does not necessarily reflect actual past financial arrangements. For example, the 2009 database, which includes data for the entire time series 1975 through 2009, no longer reflects the fact that Chrysler was combined with Daimler for several years.

In some tables and figures in this report, a single *laboratory* composite (combined city/highway) value is also shown. Because the underlying methodology for generating and reporting laboratory values has not changed since this series began in the mid-1970s, these laboratory values provide an excellent basis for comparing long-term CO_2 emissions and fuel economy trends from the perspective of vehicle design, apart from the factors that affect real-world CO_2 and fuel economy that are reflected in the adjusted values. For 2005 and later model years, laboratory composite fuel economy values are, on average, about 25 percent greater than adjusted composite fuel economy values, and laboratory composite CO_2 emissions values are, on average, about 20 percent lower than adjusted composite CO_2 values.

Formal Corporate Average Fuel Economy (CAFE) compliance data as reported by the Department of Transportation's National Highway Traffic Safety Administration (NHTSA) do not correlate precisely with either the adjusted or laboratory fuel economy values in this report. While EPA's laboratory composite fuel economy data form the cornerstone of the CAFE compliance database, NHTSA must also include credits for alternative fuel vehicles and test procedure adjustments (for cars only) in the official CAFE calculations. Accordingly, NHTSA CAFE values are at least 25 percent higher than EPA adjusted fuel economy values for model years 2005 through 2009.

In general, car/truck classifications in this database parallel classifications made by NHTSA for CAFE purposes and EPA for vehicle emissions standards. However, this report relies on engineering judgment, and typically there are a few cases each model year where the methodology used for classifying vehicles for this report results in differences in the determination of whether a given vehicle is classified as a car or a light truck. See Appendix A for a list of these exceptions.

The data presented in this report were tabulated on a model year basis, but some of the figures in this report use three-year moving averages that effectively smooth the trends, and these three-year moving averages are tabulated at the midpoint. For example, the midpoint for model years 2007, 2008, and 2009 is MY2008. Figures are based on annual data unless otherwise noted.

All of the data in this report are from vehicles certified to operate on gasoline or diesel fuel, from laboratory testing with test fuels as defined in EPA test protocols. There are no data from the very small number of vehicles that are certified to operate only on alternative fuels. The data from ethanol flexible fuel vehicles, which can operate on both an 85 percent ethanol/15 percent gasoline blend or gasoline or any mixture in between, are from gasoline operation.

While CO_2 emissions values can be arithmetically averaged, all average fuel economy values were calculated using harmonic rather than arithmetical averaging, in order to maintain mathematical integrity. See Appendix A.

The EPA database used to generate the CO_2 emissions and fuel economy values in this report was frozen in April 2009, yielding additional data beyond that used in last year's report for model years beginning in 2006, although additional data for MY2008 was added in June 2009.

Through MY2007, the CO_2 emissions, fuel economy, vehicle characteristics, and vehicle production volume data used for this report were from the formal end-of-year submissions from automakers obtained from EPA's fuel economy database that is used for CAFE compliance purposes. Accordingly, values for all model years up to 2007 can be considered final.

For MY2008, the data used in this report are based almost exclusively on formal end-of-year CAFE submissions by automakers. Accordingly, the MY2008 data are essentially final and EPA has a very high level of confidence in the data for MY2008. It is noteworthy that the 21.0 mpg adjusted fuel economy value for MY2008 in this report is 0.2 mpg higher than the projected 20.8 mpg adjusted fuel economy value for MY2008 in the 2008 report. This suggests that higher gasoline prices have led to actual 2008 production volumes that differ from the projected 2008 production levels provided to EPA by automakers in 2007.

For MY2009, EPA has exclusively used confidential pre-model year production volume projections. Accordingly, MY2009 projections are much more uncertain, particularly given the changes in the automotive marketplace driven by the economic recession and volatile fuel prices. For model years 1998 through 2006, the final laboratory fuel economy values for a given model year have varied from 0.4 mpg lower to 0.4 mpg higher compared to original estimates for the same model year that were based exclusively on projected production levels.

In the various appendices to this report, when there is no entry under "Model Year," that means there was no production volume for the data in question.

November 2009

For More Information

Light-Duty Automotive Technology, Carbon Dioxide Emissions, and Fuel Economy Trends: 1975 through 2009 (EPA420-R-09-014) is available on the Office of Transportation and Air Quality's (OTAQ) Web site at:

www.epa.gov/otaq/fetrends.htm

Printed copies are available from the OTAQ library at:

U.S. Environmental Protection Agency
Office of Transportation and Air Quality Library
2000 Traverwood Drive
Ann Arbor, MI 48105
(734) 214-4311

A copy of the *Fuel Economy Guide* giving city and highway fuel economy data for individual models is available at:

www.fueleconomy.gov

or by calling the U.S. Department of Energy at (800) 423-1363.

EPA's *Green Vehicle Guide* providing information about the air pollution emissions and fuel economy performance of individual models is available on EPA's web site at:

www.epa.gov/greenvehicles

For information about the Department of Transportation (DOT) Corporate Average Fuel Economy (CAFE) program, including a program overview, related rulemaking activities, and summaries of the fuel economy performance of individual manufacturers since 1978, see:

www.nhtsa.dot.gov and click on "Fuel Economy"

II. Introduction

Light-duty automotive technology, carbon dioxide (CO_2) emissions, and fuel economy trends are examined here, using the latest and most complete EPA data available. Past reports in this series [1-35] [1] have presented fuel economy and technology trends only, and did not include CO_2 emissions data. Section IV of this report provides a few key CO_2 emissions summary tables as well as a methodology with which a reader can convert fuel economy values from other sections of this report to equivalent CO_2 emissions levels.

When comparing data in this and previous reports, please note that revisions are made for some prior model years for which more complete and accurate production and fuel economy data have become available. In addition, changes have been made periodically in the way EPA calculates adjusted fuel economy values which means it is not appropriate to compare adjusted fuel economy values from this report with others in this series. Finally, the grouping of individual manufacturers into broader marketing groups also changes over time to reflect changes in the financial arrangements within the automobile industry.

The EPA CO_2 emissions and fuel economy database used to generate the fuel economy trends database in this report was frozen in April 2009, yielding additional data beyond that used in last year's report for model years 2006 through 2009, though additional data for MY2008 was added in June 2009.

Through MY2008, the CO_2 emissions, fuel economy, vehicle characteristics, and production volume data used for this report were from the formal end-of-year submissions from automakers obtained from EPA's database that is used for CAFE compliance purposes. For MY2009, EPA has exclusively used confidential pre-model year production projections submitted to EPA by automakers.

Accordingly, values for all model years up to 2008 can be considered final. MY2009 projections are much more uncertain, particularly given the changes in the automotive marketplace driven by the economic recession and volatile fuel prices at the time the production projections were submitted to EPA by automakers in the spring and summer of 2008. Over the last several years, the final fuel economy values for a given model year have varied from 0.4 mpg lower to 0.6 mpg higher compared to original estimates for the same model year that were based exclusively on projected production volumes.

All CO_2 emissions values in this report are production-weighted arithmetic averages and all fuel economy averages are production-weighted harmonic averages (necessary to maintain mathematical integrity). In prior reports in this series, up to and including the one for MY2000, the only fuel economy values used in this series were the laboratory-based city, highway, and composite (combined city/highway) mpg values - the same ones that are used as the basis for compliance with the fuel economy standards and the gas guzzler tax. Since the laboratory mpg values tend to over predict the mpg achieved in actual use, adjusted mpg values are used for the Government's fuel economy information programs: the *Fuel Economy Guide* and the *Fuel Economy Labels* that are on new vehicles and in EPA's *Green Vehicle Guide*.

Starting with the report issued for MY2001, this series of reports has provided fuel economy trends in adjusted mpg values in addition to the laboratory mpg values. In this way, the fuel economy trends can be shown for both laboratory mpg and mpg values which can be considered to be an estimate of on-road mpg. In the tables, these two mpg values are called "Laboratory MPG" and "Adjusted MPG," and abbreviated "LAB" MPG and "ADJ" MPG. These same metrics are used for CO_2 emissions values as well.

Where only one CO_2 or mpg value is presented in this report and it is not explicitly identified otherwise, it is the "adjusted composite" value. This value represents a combined city/highway CO_2 or fuel economy value, and is based on equations (see Appendix A) that allow a computation of adjusted city and highway values based on laboratory city and highway test values.

It is important to note that EPA revised the methodology by which EPA estimates real-world fuel economy values in December 2006. This is the third report in this series to reflect this new real-world fuel

---

[1] Numbers in brackets denote references listed in the references section of this report.

EPA-420-R-09-014

economy methodology, and every adjusted (ADJ) fuel economy value in this report for 1986 and later model years is lower than given in pre-2007 reports in this series. Accordingly, adjusted fuel economy values for 1986 and later model years should not be compared with corresponding values from older reports. These new downward adjustments are phased in, linearly, beginning in 1986, and for 2005 and later model years the new adjusted composite values are, on average, about six percent lower than under the methodology previously used by EPA. This same methodology is used to generate adjusted CO_2 emissions values as well. See Appendix A for more in-depth discussion of this new methodology and how it affects both the adjusted CO_2 and fuel economy values for individual models and the historical trends database.

The data presented in this report were tabulated on a model year basis, but many of the figures in this report use three-year moving averages which effectively smooth the trends, and these three-year moving averages are tabulated at their midpoint. For example, the midpoint for model years 2007, 2008, and 2009 is model year 2008 (See Table A-2, Appendix A). Use of the three-year moving averages results in an improvement in distinguishing real trends from what might be relatively small year-to-year variations in the data.

To facilitate comparison with data in older reports in this series, most data tables include laboratory 55/45 fuel economy values as well as the adjusted city, highway, and composite fuel economy values. Presenting both types of mpg values facilitates the use of this report by those who study either type of fuel economy metric.

The fuel economy values reported by the Department of Transportation (DOT) for compliance with the Corporate Average Fuel Economy (CAFE) compliance purposes are higher than the data in this report for four reasons:

1. The DOT data does not include the EPA real world fuel economy adjustment factors for city and highway mpg;

2. The DOT data include CAFE credits for those manufacturers that produce dedicated alternative fuel vehicles and flexible fuel vehicles (credits generated through the production of flexible fuel vehicles are currently capped at 1.2 mpg per fleet);

3. The DOT data include credits for test procedure adjustments for cars; and

4. There are a few differences in the way vehicles are classified as cars and trucks for this report compared to the way they are classified by DOT.

Accordingly, the fuel economy values in this series of reports are always lower than those reported by DOT. Table A-6, Appendix A, compares CAFE data reported by DOT with EPA adjusted and laboratory fuel economy data for MY1975-2009. Table A-7 shows a more detailed comparison for MY2008, by marketing group, of values for EPA laboratory fuel economy, alternative fuel vehicle credits, test procedure adjustment credits for cars, and NHTSA CAFE performance (the latter based on mid-model year estimates).

EPA-420-R-09-014
November 2009

Other Variables

All vehicle weight data are based on inertia weight class (nominally curb weight plus 300 pounds). For vehicles with inertia weights up to and including the 3000-pound inertia weight class, these classes have 250-pound increments. For vehicles above the 3000-pound inertia weight class (i.e., vehicles 3500 pounds and above), 500-pound increments are used.

All interior volume data for cars built after model year 1977 are based on the metric used to classify cars for the DOE/EPA *Fuel Economy Guide*. The car interior volume combines the passenger compartment and trunk/cargo space. In the *Fuel Economy Guide*, interior volume is undefined for the two-seater class; for this series of reports, all two-seater cars have been assigned an interior volume value of 50 cubic feet.

The light truck data used in this series of reports includes only vehicles classified as light trucks with gross vehicle weight ratings (GVWR) up to 8500 pounds (lb). Vehicles with GVWR above 8500 lb are not included in the database used for this report. Omitting these vehicles influences the overall averages for all variables studied in this report. The most recent estimates we have made for the impact of these greater than 8500 lb GVWR vehicles was made for model year 2001. In that year, there were roughly 931,000 vehicles above 8500 lb GVWR. A substantial fraction (42 percent) of the MY2001 vehicles above 8500 lb GVWR were powered by diesel engines, and three-fourths of the vehicles over 8500 lb GVWR were pickup trucks. Adding in the trucks above 8500 lb GVWR would have increased the truck market share for that year by three percentage points.

Based on a limited amount of actual laboratory fuel economy data, MY2001 trucks with GVWR greater than 8500 lb GVWR are estimated to have fuel economy values about 14 percent lower than the average of trucks below 8500 lb GVWR. The combined fleet of all vehicles under 8500 lb GVWR and trucks over 8500 lb GVWR is estimated to average a few percent less in fuel economy compared to that for just the vehicles with less than 8500 lb GVWR.

In addition to mpg, some tables in this report contain alternate measures of vehicle fuel efficiency as used in reference 17.

"Ton-MPG" is defined as a vehicle's mpg multiplied by its inertia weight in tons. Ton-MPG is a measure of powertrain/drive-line efficiency. Just as an increase in vehicle mpg at constant weight can be considered an improvement in a vehicle's efficiency, an increase in a vehicle's weight at constant mpg can also be considered an improvement. "CO_2/ton" is the equivalent CO_2 metric and is reported in Section IV.

"Cubic-feet-MPG" for cars is defined in this report as the product of a car's mpg and its interior volume, including trunk space. This metric associates a relative measure of a vehicle's ability to transport both passengers and their cargo. An increase in vehicle volume at constant mpg could be considered an improvement just as an increase in mpg at constant volume can be. "CO_2/cubic feet" values are given in Section IV.

"Cubic-feet-ton-MPG" is defined in this report as a combination of the two previous metrics, i.e., a car's mpg multiplied by its weight in tons and also by its interior volume. It ascribes vehicle utility to fuel economy, weight and volume. "CO_2/ton-cubic feet"" is the equivalent CO_2 metric and is shown in Section IV.

This report also includes an estimate of 0-to-60 mph acceleration time, calculated from engine rated horsepower and vehicle inertia weight, from the relationship:

$$t = F \ (HP/WT)^{-f}$$

where the values used for F and f coefficients are .892 and .805 respectively for vehicles with automatic transmissions and .967 and .775 respectively for those with manual transmissions [36]. Other authors [37, 38, and 39] have evaluated the relationships between weight, horsepower, and 0-to-60 acceleration time and have calculated and published slightly different values for the F and f coefficients. Since the equation form and coefficients were developed for vehicles with conventional powertrains with gasoline-fueled engines, we have not used the equation to estimate 0-to-60 time for vehicles with hybrid powertrains or diesel engines. Published values are used for these vehicles instead.

The 0-to-60 estimate used in this report is intended to provide a quantitative time "index" of vehicle performance capability. It is the authors' engineering judgment that, given the differences in test methods for measuring 0-to-60 time and given the fact that the weight is based on inertia weight, use of these other published values for the F and f coefficients would not result in statistically significantly different 0-to-60 averages or trends. The results of a similar calculation of estimated "top speed" are also included in some tables.

Grouping all vehicles into classes and then constructing time trends can provide interesting and useful results. These results, however, are a strong function of the class definitions. Classes based on other definitions than those used in this report are possible, and results from these other classifications may also be useful.

For cars, vehicle classification as to vehicle type, size class, and manufacturer/origin generally follows fuel economy label, *Fuel Economy Guide*, and fuel economy standards protocols; exceptions are listed in Table A-3, Appendix A. In many of the passenger car tables, large sedans and wagons are aggregated as "Large," midsize sedans and wagons are aggregated as "Midsize," and "Small" includes all other cars. In some of the car tables, an alternative classification system is used, namely: Large Cars, Large Wagons, Midsize Cars, Midsize Wagons, Small Cars, and Small Wagons with the EPA Two-Seater, Mini-Compact, Subcompact, and Compact car classes are combined into the "Small Car" class. In some of the tables and figures in this report, only four vehicle types are used. In these cases, wagons have been merged with cars. This is because the wagon production fraction for some instances is so small that the information is more conveniently represented by combining the two vehicle types. When they have been combined, the differences between them are not important.

The truck classification scheme used for all model years in this report is slightly different from that used in some previous reports in this series, because pickups, vans, and sports utility vehicles (SUVs) are sometimes each subdivided as "Small," "Midsize," and "Large." These truck size classifications are based primarily on published wheelbase data according to the following criteria:

	Pickup	Van	SUV
Small	Less than 105"	Less than 109"	Less than 100"
Midsize	105" to 115"	109" to 124"	100" to 110"
Large	More than 115"	More than 124"	More than 110"

This classification scheme is similar to that used in many trade and consumer publications. For those vehicle nameplates with a variety of wheelbases, the size classification was determined by considering only the smallest wheelbase produced. The classification of a vehicle for this report is based on the authors' engineering judgment and is not a replacement for definitions used in implementing automotive standards legislation.

Published data is also used for three other vehicle characteristics for which data is not currently being submitted to EPA by the automotive manufacturers, or to supplement data that is submitted to EPA: (1) engines with variable valve timing (VVT) that use either cams or electric solenoids to provide variable intake and/or exhaust valve timing and in some cases valve lift; (2) engines with cylinder deactivation, which involves allowing the valves of selected cylinders of the engine to remain closed under certain driving conditions; and (3) vehicle footprint, which is the product of wheelbase times average track width and upon which future CAFE standards, and likely future CO_2 emissions standards, will be based..

EPA-420-R-09-014   November 2009

III. Fuel Economy Trends

Figure 1 and Table 1 depict time trends in car, light truck, and car-plus-light truck fuel economy. Also shown on Figure 1 is the fraction of the combined fleet that are light trucks and trend lines representing three-year moving averages of the fuel economy and truck production fraction data. Since 1975, the fuel economy of the combined car and light truck fleet has moved through several phases:

1. A rapid increase from 1975 through 1981;

2. A slow increase until reaching its peak in 1987;

3. A gradual decline until 2004; and

4. An increase beginning in 2005.

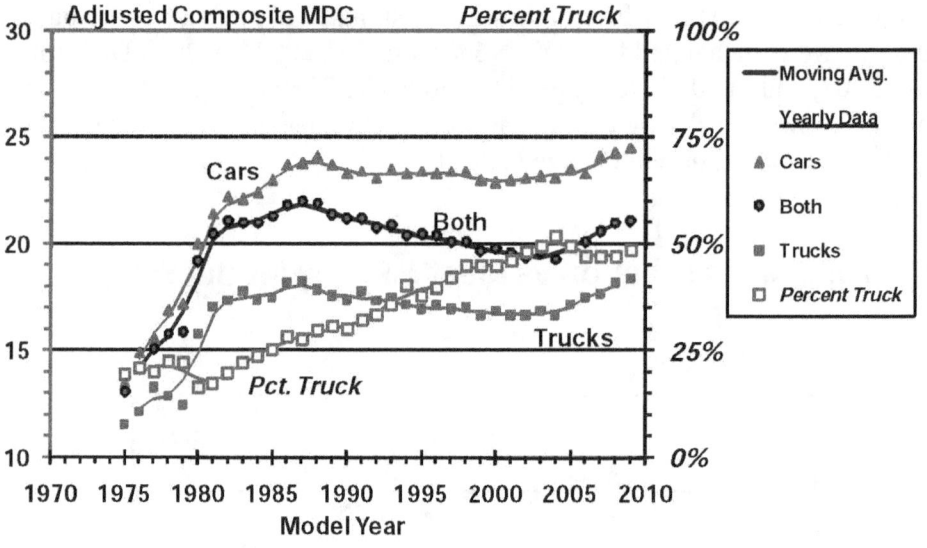

Figure 1

EPA-420-R-09-014   5   November 2009

As shown in Table 1, the projected MY2009 fleetwide fuel economy value of 21.1 mpg is the highest value since 1991 and is 1.8 mpg higher than the 2004 value of 19.3 mpg, which was the lowest value since 1980. Projected industry-wide MY2009 production is not shown in Table 1, as it is expected that actual MY2009 production will be 30 to 40 percent lower than automaker projections to EPA in spring/summer 2008. Average fleetwide fuel economy has now increased for five consecutive years. These increases reverse the longer term trend of declining adjusted composite fuel economy since its peak in 1987. Most of the increase in overall fuel economy since 2004 is due to higher truck fuel economy (likely due at least in part to higher truck CAFE standards in recent years), as truck fuel economy has increased by 1.7 mpg since 2004, while car fuel economy has increased by 1.4 mpg. The 21.1 mpg adjusted fuel economy value projected for 2009 is 0.9 mpg below the peak in 1987, but this difference is due to the new methodology for calculating adjusted fuel economy values that is phased in over the 1986 – 2005 timeframe. As shown in Table 1, based on laboratory 55/45 fuel economy values which are based on vehicle design considerations only, the projected fleetwide fuel economy value of 26.4 mpg is an all-time record, and is 0.5 mpg higher than the previous peak of 25.9 mpg in 1987.

Figure 1 shows that the estimated light truck share of the market, based on the three-year moving average trend, has leveled off at about 50 percent. Figure 2 compares laboratory 55/45 fuel economy for the combined car and truck fleet and the production fraction for trucks.

The MY2009 adjusted fuel economy for cars is estimated to average 24.5 mpg, which is an all-time high. For MY2009, the adjusted fuel economy for light trucks is estimated to average 18.4 mpg, also a record high. Fuel economy standards were unchanged for MY1996 through MY2004. In 2003 DOT raised the truck CAFE standards for 2005 – 2007, and in 2006 DOT raised the truck CAFE standards for 2008 – 2011. The recent fuel economy improvement for trucks is likely due, in part, to these higher standards. The CAFE standard for cars has not been changed since 1990, but will change in 2011.

Truck Sales Fraction vs Fleet MPG by Model Year

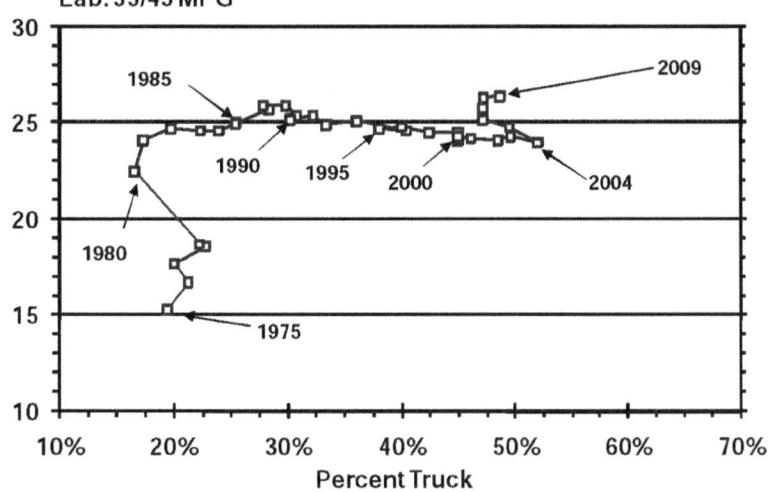

Figure 2

Table 1

Fuel Economy Characteristics of 1975 to 2009 Light Duty Vehicles

Cars

MODEL YEAR	PROD (000)	FRAC	<---------- FUEL ECONOMY ---------->						TON -MPG	CU-FT -MPG	CU-FT- TON-MPG
			LAB CITY	LAB HWY	LAB 55/45	ADJ CITY	ADJ HWY	ADJ COMP			
1975	8237	.806	13.7	19.5	15.8	12.3	15.2	13.5	27.6		
1976	9722	.788	15.2	21.3	17.5	13.7	16.6	14.9	30.2		
1977	11300	.800	16.0	22.3	18.3	14.4	17.4	15.6	31.0	1780	3423
1978	11175	.773	17.2	24.5	19.9	15.5	19.1	16.9	30.6	1908	3345
1979	10794	.778	17.7	24.6	20.3	15.9	19.2	17.2	30.2	1922	3301
1980	9443	.835	20.3	29.0	23.5	18.3	22.6	20.0	31.2	2136	3273
1981	8733	.827	21.7	31.1	25.1	19.6	24.2	21.4	33.1	2338	3547
1982	7819	.803	22.3	32.7	26.0	20.1	25.5	22.2	34.2	2419	3645
1983	8002	.777	22.1	32.7	25.9	19.9	25.5	22.1	34.7	2476	3776
1984	10675	.761	22.4	33.3	26.3	20.2	26.0	22.4	35.1	2482	3776
1985	10791	.746	23.0	34.3	27.0	20.7	26.8	23.0	35.8	2553	3884
1986	11015	.717	23.7	35.5	27.9	21.2	27.6	23.7	36.2	2598	3899
1987	10731	.722	23.9	35.9	28.1	21.2	27.7	23.8	36.2	2584	3872
1988	10736	.702	24.2	36.6	28.6	21.4	28.2	24.1	36.9	2631	3963
1989	10018	.693	23.8	36.3	28.1	20.9	27.9	23.7	36.8	2591	3977
1990	8810	.698	23.4	36.0	27.8	20.5	27.5	23.3	37.1	2528	3984
1991	8524	.678	23.6	36.3	28.0	20.5	27.6	23.4	37.0	2540	3970
1992	8108	.666	23.1	36.3	27.6	20.0	27.5	23.1	37.4	2534	4071
1993	8456	.640	23.6	36.9	28.2	20.3	27.9	23.5	37.7	2580	4098
1994	8415	.596	23.4	36.9	28.0	20.0	27.7	23.3	37.9	2554	4108
1995	9396	.620	23.6	37.6	28.3	20.0	28.1	23.4	38.3	2584	4171
1996	7890	.600	23.5	37.6	28.3	19.8	28.0	23.3	38.3	2572	4186
1997	8335	.576	23.7	37.7	28.4	19.8	28.0	23.4	38.3	2565	4168
1998	7972	.551	23.7	37.9	28.5	19.7	28.0	23.4	38.7	2565	4210
1999	8379	.551	23.4	37.4	28.2	19.4	27.5	23.0	38.7	2531	4237
2000	9128	.551	23.5	37.3	28.2	19.3	27.3	22.9	38.6	2534	4246
2001	8408	.539	23.7	37.6	28.4	19.4	27.3	23.0	39.1	2551	4280
2002	8304	.515	24.0	37.6	28.6	19.4	27.2	23.1	39.3	2561	4311
2003	7951	.504	24.2	38.1	28.9	19.5	27.5	23.2	40.0	2582	4378
2004	7538	.480	24.1	38.2	28.9	19.3	27.4	23.1	40.3	2601	4464
2005	8027	.505	24.7	38.7	29.5	19.6	27.6	23.5	41.0	2677	4590
2006	7993	.529	24.4	38.5	29.2	19.4	27.5	23.3	41.6	2655	4649
2007	8085	.529	25.4	39.7	30.3	20.1	28.3	24.1	42.8	2733	4734
2008	7345	.528	25.6	40.0	30.5	20.3	28.5	24.3	43.3	2749	4784
2009	----	.513	25.9	40.4	30.9	20.5	28.8	24.5	43.8	2786	4858

EPA-420-R-09-014

Table 1   (Continued)

Fuel Economy Characteristics of 1975 to 2009 Light Duty Vehicles

Trucks

MODEL YEAR	PROD (000)	FRAC	LAB CITY	LAB HWY	LAB 55/45	ADJ CITY	ADJ HWY	ADJ COMP	TON -MPG
1975	1987	.194	12.1	16.2	13.7	10.9	12.7	11.6	24.2
1976	2612	.212	12.8	16.9	14.4	11.5	13.2	12.2	26.0
1977	2823	.200	14.0	18.1	15.6	12.6	14.1	13.3	28.0
1978	3273	.227	13.8	17.5	15.2	12.4	13.7	12.9	27.5
1979	3088	.222	13.4	16.8	14.7	12.1	13.1	12.5	27.3
1980	1863	.165	16.5	21.9	18.6	14.8	17.1	15.8	30.9
1981	1821	.173	17.8	23.9	20.1	16.0	18.6	17.1	33.0
1982	1914	.197	18.1	24.4	20.5	16.3	19.0	17.4	33.7
1983	2300	.223	18.3	25.2	20.9	16.5	19.6	17.8	34.0
1984	3345	.239	17.9	24.8	20.5	16.1	19.3	17.4	33.5
1985	3669	.254	18.0	24.9	20.6	16.2	19.4	17.5	33.7
1986	4350	.283	18.8	25.9	21.4	16.8	20.2	18.2	34.3
1987	4134	.278	18.8	26.5	21.6	16.8	20.5	18.3	34.2
1988	4559	.298	18.3	26.2	21.2	16.2	20.2	17.9	34.5
1989	4435	.307	18.1	25.8	20.9	15.9	19.8	17.6	34.7
1990	3805	.302	17.8	25.9	20.7	15.6	19.8	17.4	35.1
1991	4049	.322	18.3	26.6	21.3	15.9	20.3	17.8	35.3
1992	4064	.334	17.8	26.2	20.8	15.5	19.9	17.4	35.4
1993	4754	.360	17.9	26.5	21.0	15.5	20.1	17.5	35.7
1994	5710	.404	17.8	26.1	20.8	15.3	19.7	17.2	35.7
1995	5749	.380	17.5	25.9	20.5	15.0	19.5	17.0	35.7
1996	5254	.400	17.7	26.5	20.8	15.1	19.9	17.2	36.6
1997	6124	.424	17.6	26.1	20.6	14.8	19.5	17.0	36.9
1998	6485	.449	17.7	26.6	20.9	14.9	19.8	17.1	36.8
1999	6839	.449	17.4	26.0	20.5	14.6	19.2	16.7	37.0
2000	7447	.449	17.7	26.2	20.8	14.7	19.4	16.9	37.1
2001	7202	.461	17.6	26.0	20.6	14.6	19.1	16.7	37.4
2002	7815	.485	17.6	26.0	20.6	14.4	19.1	16.7	38.0
2003	7824	.496	17.8	26.5	20.9	14.6	19.3	16.9	38.7
2004	8173	.520	17.7	26.5	20.8	14.3	19.2	16.7	39.4
2005	7866	.495	18.2	27.4	21.4	14.6	19.8	17.2	40.2
2006	7111	.471	18.5	27.8	21.8	14.9	20.1	17.5	40.9
2007	7192	.471	18.7	28.3	22.1	15.1	20.4	17.7	42.1
2008	6554	.472	19.2	29.1	22.7	15.5	21.0	18.2	43.0
2009	----	.487	19.4	29.6	22.9	15.6	21.4	18.4	43.5

EPA-420-R-09-014

Table 1   (Continued)

Fuel Economy Characteristics of 1975 to 2009 Light Duty Vehicles

Cars and Trucks

MODEL YEAR	PROD (000)	FRAC	<---------- FUEL ECONOMY ---------->						TON -MPG
			LAB CITY	LAB HWY	LAB 55/45	ADJ CITY	ADJ HWY	ADJ COMP	
1975	10224	1.000	13.4	18.7	15.3	12.0	14.6	13.1	26.9
1976	12334	1.000	14.6	20.2	16.7	13.2	15.7	14.2	29.3
1977	14123	1.000	15.6	21.3	17.7	14.0	16.6	15.1	30.4
1978	14448	1.000	16.3	22.5	18.6	14.7	17.5	15.8	29.9
1979	13882	1.000	16.5	22.3	18.7	14.9	17.4	15.9	29.5
1980	11306	1.000	19.6	27.5	22.5	17.6	21.5	19.2	31.2
1981	10554	1.000	20.9	29.5	24.1	18.8	23.0	20.5	33.1
1982	9732	1.000	21.3	30.7	24.7	19.2	23.9	21.1	34.1
1983	10302	1.000	21.2	30.6	24.6	19.0	23.9	21.0	34.5
1984	14020	1.000	21.2	30.8	24.6	19.1	24.0	21.0	34.7
1985	14460	1.000	21.5	31.3	25.0	19.3	24.4	21.3	35.3
1986	15365	1.000	22.1	32.2	25.7	19.8	25.0	21.8	35.7
1987	14865	1.000	22.2	32.6	25.9	19.8	25.3	22.0	35.7
1988	15295	1.000	22.1	32.7	25.9	19.6	25.2	21.9	36.2
1989	14453	1.000	21.7	32.3	25.4	19.1	24.8	21.4	36.2
1990	12615	1.000	21.4	32.2	25.2	18.7	24.6	21.2	36.5
1991	12573	1.000	21.6	32.5	25.4	18.8	24.7	21.2	36.5
1992	12172	1.000	21.0	32.1	24.9	18.2	24.4	20.8	36.8
1993	13211	1.000	21.2	32.4	25.1	18.2	24.4	20.9	37.0
1994	14125	1.000	20.8	31.6	24.6	17.8	23.8	20.4	37.0
1995	15145	1.000	20.8	32.1	24.7	17.7	24.1	20.5	37.3
1996	13144	1.000	20.8	32.2	24.8	17.6	24.0	20.4	37.6
1997	14459	1.000	20.6	31.8	24.5	17.4	23.6	20.1	37.7
1998	14458	1.000	20.6	31.9	24.5	17.2	23.6	20.1	37.9
1999	15218	1.000	20.3	31.2	24.1	16.9	23.0	19.7	38.0
2000	16574	1.000	20.5	31.4	24.3	16.9	23.0	19.8	37.9
2001	15610	1.000	20.5	31.1	24.2	16.8	22.8	19.6	38.3
2002	16119	1.000	20.4	30.9	24.1	16.6	22.5	19.4	38.7
2003	15775	1.000	20.6	31.3	24.3	16.7	22.7	19.6	39.4
2004	15711	1.000	20.2	31.0	24.0	16.3	22.4	19.3	39.9
2005	15893	1.000	21.0	32.1	24.8	16.8	23.1	19.9	40.6
2006	15105	1.000	21.2	32.6	25.2	17.0	23.4	20.1	41.2
2007	15277	1.000	21.8	33.4	25.8	17.4	24.0	20.6	42.5
2008	13900	1.000	22.1	34.0	26.3	17.7	24.4	21.0	43.2
2009	----	1.000	22.2	34.3	26.4	17.8	24.6	21.1	43.6

The distribution of fuel economy in any model year is of interest. In Figure 3, highlights of the distribution of car mpg are shown. Since 1975, half of the cars have consistently been within a few mpg of each other. The fuel economy difference between the least efficient and most efficient car increased from about 20 mpg in 1975 to nearly 50 mpg in 1986, but was less than 35 mpg in 1999. With the introduction for sale of the Honda Insight gasoline-electric hybrid vehicle in MY2000, the range once again approached 50 mpg. The increased market share of hybrid cars also accounts for the increase in the fuel economy of the best one percent of cars with the cutpoint for this stratum now over 40 mpg. The ratio of the highest to lowest has increased from about three to one in 1975 to nearly five to one today, because the fuel economy of the least fuel efficient cars has remained roughly constant in comparison to the most fuel efficient cars whose fuel economy has more than doubled.

The overall fuel economy distribution trend for trucks (see Figure 4) is narrower than that for cars, with a peak in the efficiency of the most efficient truck in the early 1980s when small pickup trucks equipped with diesel engines were being sold. As a result, the fuel economy range between the most efficient and least efficient truck peaked at about 25 mpg in 1982. The fuel economy range for trucks then narrowed, but with the introduction of the hybrid Escape SUV in MY2005, it is now over 20 mpg. Like cars, half of the trucks built each year have always been within a few mpg of each year's average fuel economy value. Appendix C contains additional fuel economy distribution data.

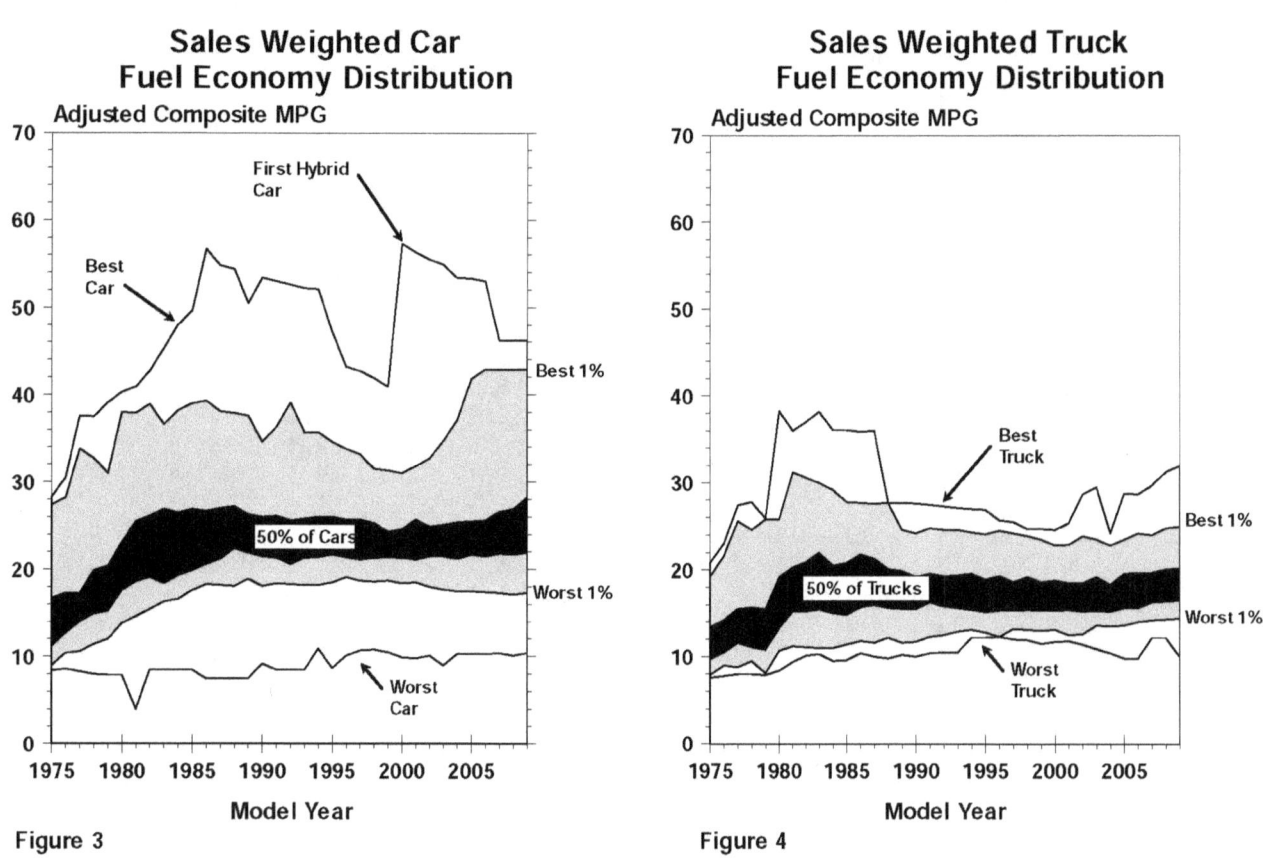

Figure 3

Figure 4

EPA-420-R-09-014   10   November 2009

Table 2

Vehicle Size and Design Characteristics of 1975 to 2009

Cars

										Percent By:		
			Vehicle Characteristics:									
MODEL YEAR	PROD FRAC	ADJ COMP MPG	VOL CU-FT	WGHT LB	FOOT PRNT SQFT	ENG HP	HP/ WT	0-60 TIME	TOP SPD	VEHICLE SIZE		
										SMALL	MID	LARGE
1975	.806	13.5		4058		136	.0331	14.2	111	55.4	23.3	21.3
1976	.788	14.9		4059		134	.0324	14.4	110	55.4	25.2	19.4
1977	.800	15.6	110	3944		133	.0335	14.0	111	51.9	24.5	23.5
1978	.773	16.9	109	3588		124	.0342	13.7	111	44.7	34.4	21.0
1979	.778	17.2	109	3485		119	.0338	13.8	110	43.7	34.2	22.1
1980	.835	20.0	104	3101		100	.0322	14.3	107	54.4	34.4	11.3
1981	.827	21.4	106	3076		99	.0320	14.4	106	51.5	36.4	12.2
1982	.803	22.2	106	3054		99	.0320	14.4	106	56.5	31.0	12.5
1983	.777	22.1	109	3112		104	.0330	14.0	108	53.1	31.8	15.1
1984	.761	22.4	108	3099		106	.0339	13.8	109	57.4	29.4	13.2
1985	.746	23.0	108	3093		111	.0355	13.3	111	55.7	28.9	15.4
1986	.717	23.7	107	3041		111	.0360	13.2	111	59.5	27.9	12.6
1987	.722	23.8	107	3031		112	.0365	13.0	112	63.5	24.3	12.2
1988	.702	24.1	107	3047		116	.0375	12.8	113	64.8	22.3	12.8
1989	.693	23.7	108	3099		121	.0387	12.5	115	58.3	28.2	13.5
1990	.698	23.3	107	3176		129	.0401	12.1	117	58.6	28.7	12.8
1991	.678	23.4	107	3154		132	.0413	11.8	118	61.5	26.2	12.3
1992	.666	23.1	108	3240		141	.0428	11.5	120	56.5	27.8	15.6
1993	.640	23.5	108	3207		138	.0425	11.6	120	57.2	29.5	13.3
1994	.596	23.3	108	3250		143	.0432	11.4	121	58.5	26.1	15.4
1995	.620	23.4	109	3263		152	.0460	10.9	125	57.3	28.6	14.0
1996	.600	23.3	109	3282		154	.0464	10.8	125	54.3	32.0	13.6
1997	.576	23.4	109	3274		156	.0469	10.7	126	55.1	30.6	14.3
1998	.551	23.4	109	3306		159	.0475	10.6	127	49.4	39.1	11.4
1999	.551	23.0	109	3365		164	.0481	10.5	128	47.7	39.7	12.6
2000	.551	22.9	110	3369		168	.0492	10.4	129	47.5	34.3	18.2
2001	.539	23.0	109	3380		168	.0492	10.3	129	50.9	32.3	16.8
2002	.515	23.1	109	3391		173	.0504	10.2	131	48.6	36.3	15.1
2003	.504	23.2	109	3421		176	.0510	10.0	132	50.8	33.4	15.9
2004	.480	23.1	110	3462		182	.0521	9.8	133	47.4	35.5	17.0
2005	.505	23.5	111	3463		182	.0518	9.8	133	44.2	38.9	16.8
2006	.529	23.3	112	3534		194	.0540	9.6	136	46.2	32.9	20.9
2007	.529	24.1	110	3507		189	.0531	9.6	135	44.6	40.0	15.4
2008	.528	24.3	110	3527	45.4	193	.0536	9.6	136	44.6	35.9	19.5
2009	.513	24.5	111	3533		198	.0548	9.5	137	43.8	33.3	23.0

Table 2 (Continued)

Vehicle Size and Design Characteristics of 1975 to 2009

Trucks

					<------- Vehicle Characteristics: ------->					<--Percent By:-->	
MODEL YEAR	PROD FRAC	ADJ COMP MPG	WGHT LB	FOOT PRNT SQFT	ENG HP	HP/ WT	0-60 TIME	TOP SPD	VAN	VEHICLE SUV	TYPE PICKUP
1975	.194	11.6	4072		142	.0349	13.6	114	23.0	9.4	67.6
1976	.212	12.2	4155		141	.0340	13.8	113	19.2	9.3	71.4
1977	.200	13.3	4135		147	.0356	13.3	115	18.2	10.0	71.8
1978	.227	12.9	4151		146	.0351	13.4	114	19.1	11.6	69.3
1979	.222	12.5	4252		138	.0325	14.3	111	15.6	13.0	71.5
1980	.165	15.8	3869		121	.0313	14.5	108	13.0	9.9	77.1
1981	.173	17.1	3806		119	.0311	14.6	108	13.5	7.5	79.1
1982	.197	17.4	3806		120	.0317	14.5	109	16.2	8.5	75.3
1983	.223	17.8	3763		118	.0313	14.5	108	16.6	12.6	70.8
1984	.239	17.4	3782		118	.0310	14.7	108	20.2	18.7	61.1
1985	.254	17.5	3795		124	.0326	14.1	110	23.3	20.0	56.6
1986	.283	18.2	3738		123	.0330	14.0	110	24.0	17.8	58.2
1987	.278	18.3	3713		131	.0351	13.3	113	26.9	21.1	51.9
1988	.298	17.9	3841		141	.0366	12.9	115	24.8	21.2	53.9
1989	.307	17.6	3921		146	.0372	12.8	116	28.8	20.9	50.3
1990	.302	17.4	4005		151	.0377	12.6	117	33.2	18.6	48.2
1991	.322	17.8	3948		150	.0379	12.6	117	25.5	27.0	47.4
1992	.334	17.4	4056		155	.0382	12.5	118	30.0	24.7	45.3
1993	.360	17.5	4073		162	.0398	12.1	120	30.3	27.6	42.1
1994	.404	17.2	4125		166	.0403	12.0	121	24.8	28.4	46.7
1995	.380	17.0	4184		168	.0401	12.0	121	28.9	31.6	39.5
1996	.400	17.2	4225		179	.0423	11.5	124	26.8	36.0	37.2
1997	.424	17.0	4344		187	.0429	11.4	126	20.7	40.0	39.3
1998	.449	17.1	4283		187	.0435	11.2	126	23.0	39.8	37.2
1999	.449	16.7	4412		197	.0446	11.0	128	21.4	41.4	37.2
2000	.449	16.9	4375		197	.0448	11.0	128	22.7	42.2	35.1
2001	.461	16.7	4463		209	.0466	10.6	131	17.1	47.9	35.0
2002	.485	16.7	4546		219	.0482	10.4	134	15.9	53.6	30.5
2003	.496	16.9	4586		221	.0481	10.4	134	15.7	52.6	31.6
2004	.520	16.7	4710		236	.0501	10.0	137	11.7	57.7	30.7
2005	.495	17.2	4668		237	.0505	10.0	137	18.8	51.9	29.2
2006	.471	17.5	4665		235	.0502	10.0	137	16.4	52.8	30.8
2007	.471	17.7	4752		248	.0520	9.8	140	11.8	58.8	29.4
2008	.472	18.2	4710	52.9	247	.0522	9.7	140	11.8	60.8	27.4
2009	.487	18.4	4712		253	.0534	9.6	142	9.3	65.8	24.9

Table 2   (Continued)

Vehicle Size and Design Characteristics of 1975 to 2009

Cars and Trucks

<--------- Vehicle Characteristics: --------->

MODEL YEAR	PROD FRAC	ADJ COMP MPG	WGHT LB	FOOT PRNT SQFT	ENG HP	HP/ WT	0-60 TIME	TOP SPD
1975	1.000	13.1	4060		137	.0335	14.1	112
1976	1.000	14.2	4079		135	.0328	14.3	111
1977	1.000	15.1	3982		136	.0339	13.8	112
1978	1.000	15.8	3715		129	.0344	13.6	112
1979	1.000	15.9	3655		124	.0335	13.9	110
1980	1.000	19.2	3228		104	.0320	14.3	107
1981	1.000	20.5	3202		102	.0318	14.4	107
1982	1.000	21.1	3202		103	.0320	14.4	107
1983	1.000	21.0	3257		107	.0327	14.1	108
1984	1.000	21.0	3262		109	.0332	14.0	109
1985	1.000	21.3	3271		114	.0347	13.5	110
1986	1.000	21.8	3238		114	.0351	13.4	111
1987	1.000	22.0	3221		118	.0361	13.1	112
1988	1.000	21.9	3283		123	.0372	12.8	114
1989	1.000	21.4	3351		129	.0382	12.5	115
1990	1.000	21.2	3426		135	.0394	12.2	117
1991	1.000	21.2	3410		138	.0402	12.1	118
1992	1.000	20.8	3512		145	.0413	11.8	120
1993	1.000	20.9	3519		147	.0416	11.8	120
1994	1.000	20.4	3603		152	.0420	11.7	121
1995	1.000	20.5	3613		158	.0438	11.3	123
1996	1.000	20.4	3659		164	.0447	11.1	125
1997	1.000	20.1	3727		169	.0452	11.0	126
1998	1.000	20.1	3744		171	.0457	10.9	126
1999	1.000	19.7	3835		179	.0465	10.7	128
2000	1.000	19.8	3821		181	.0472	10.6	129
2001	1.000	19.6	3879		187	.0480	10.5	130
2002	1.000	19.4	3951		195	.0493	10.3	132
2003	1.000	19.6	3999		199	.0496	10.2	133
2004	1.000	19.3	4111		211	.0511	9.9	135
2005	1.000	19.9	4059		209	.0512	9.9	135
2006	1.000	20.1	4067		213	.0522	9.8	137
2007	1.000	20.6	4093		217	.0525	9.7	137
2008	1.000	21.0	4085	49.0	219	.0529	9.7	138
2009	1.000	21.1	4108		225	.0541	9.5	139

Ton-MPG by Model Year
(Three Year Moving Average)

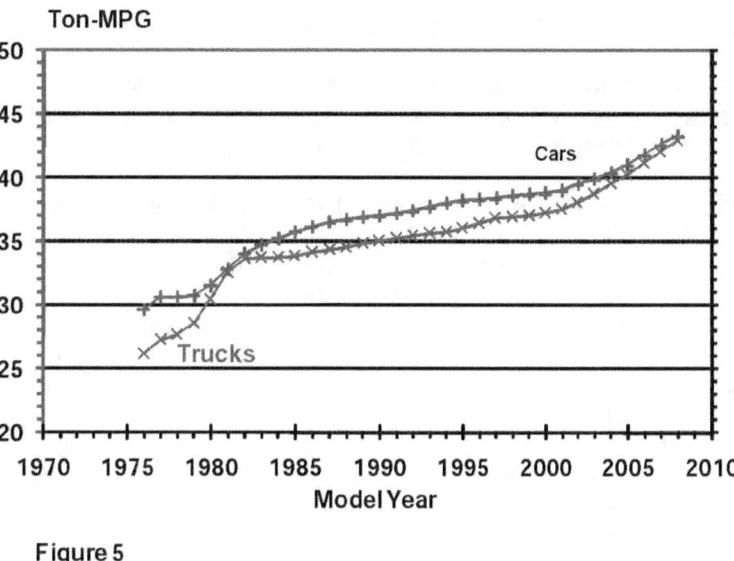

Figure 5

As shown in Table 2, the average weight of the overall fleet has remained relatively constant since 2004, with a slight increase in car weight offset by a small decrease in truck market share (as trucks have a higher average weight than cars). Overall average horsepower has continued to increase. The projected 2009 weight has increased by over 900 pounds and the average horsepower level has more than doubled since 1981.

The long term trends for both weight and performance have been steady increases. As shown in Figure 5, since 1975 Ton-MPG for both cars and trucks increased substantially; i.e., over 60% for cars and 80% for trucks. Typically, Ton-MPG for both vehicle types has increased at a rate of about one or two percent a year.

Another dramatic trend over that time frame has been the substantial increase in performance of cars and light trucks as measured by their estimated 0-to-60 time. These trends are shown graphically in Figure 6 (for cars) and Figure 7 (for light trucks) which are plots of fuel economy versus performance, with model years as indicated. Both graphs show the same story: in the late 1970s and early 1980s, responding to the regulatory requirements for mpg improvement, the industry increased mpg and kept performance roughly constant. After the regulatory mpg requirements stabilized, mpg improvements slowed and performance dramatically improved. This trend toward increased performance is as important as the truck market share trend in understanding trends in overall fleet mpg. Figures 8 and 9 are similar to Figures 6 and 7, but show the trends in weight and laboratory fuel economy and show that the era of weight reductions that took place for both cars and trucks between 1975 and the early 1980s has been followed by an era of weight increases until recently.

Table 2 also includes, for the first time, a column for vehicle footprint, in square feet. Footprint is one metric for vehicle size, and is the product of wheelbase and average track width. Essentially, footprint is the area defined by the four points where the tires touch the ground. Footprint is of interest as MY2008 – 2010 light truck CAFE standards allow manufacturers the option to choose footprint-based standards, MY2011 passenger car and light truck CAFE standards are based exclusively on footprint-mpg curves, and MY2012 – 2016 CAFE and CO_2 emissions standards may be footprint-based as well. EPA does not receive footprint data from manufacturers, so the MY2008 footprint data in Table 2 is tabulated from external sources. MY2008 is the first year for which we are reporting footprint data, but we expect to do so in future years as well. For MY2008, industry-wide footprint values were 45.4 square feet for cars, 52.9 square feet for trucks, and 49.0 square feet for cars and trucks combined.

EPA-420-R-09-014   14   November 2009

Car 55/45 Laboratory MPG vs 0 to 60 Time by Model Year

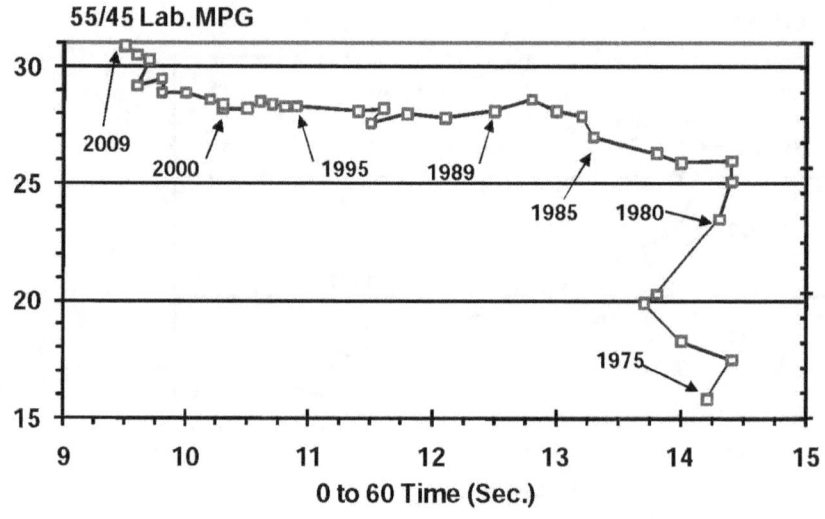

Figure 6

Truck 55/45 Laboratory MPG vs 0 to 60 Time by Model Year

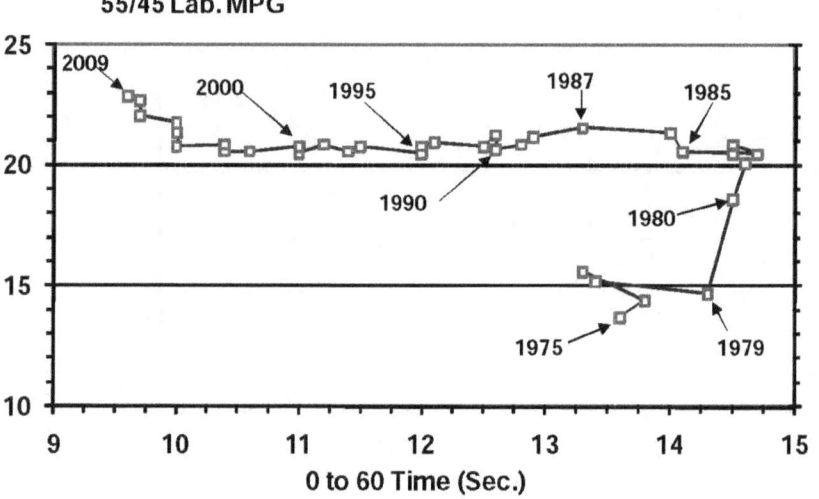

Figure 7

EPA-420-R-09-014   15   November 2009

Car 55/45 Laboratory MPG vs Inertia Weight by Model Year

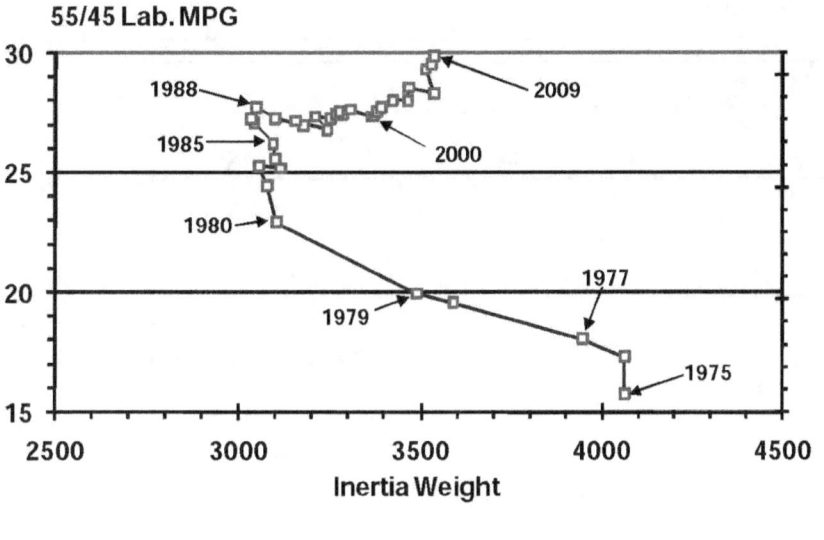

Figure 8

Truck 55/45 Laboratory MPG vs Inertia Weight by Model Year

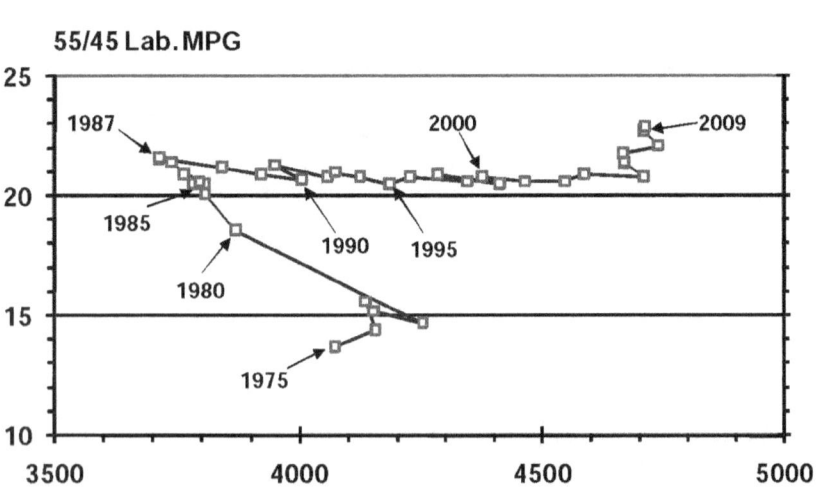

Figure 9

IV. Carbon Dioxide Emissions Trends

This new section focuses on light-duty vehicle tailpipe carbon dioxide (CO_2) emissions data that are measured over the EPA city and highway test procedures.

CO_2 is the most important greenhouse gas, responsible for a majority of all global, anthropogenic greenhouse gas emissions. Light-duty vehicles emit approximately 20 percent of total U.S. CO_2 emissions. In April 2007, the U.S. Supreme Court determined that CO_2 is a pollutant under the Clean Air Act[2], and in April 2009, EPA proposed a finding that CO_2 and other greenhouse gases from new motor vehicles and new motor vehicle engines cause or contribute to air pollution that may reasonably be anticipated to endanger public health and welfare.[3] In September 2009, EPA proposed the first-ever light-duty vehicle greenhouse gas emissions standards, under the Clean Air Act, for MY2012-2016.[4] EPA expects to finalize these standards in early 2010. These standards will be part of a new, harmonized National Policy that also includes new CAFE standards for MY2012-2016, established and administered by DOT's National Highway Traffic Safety Administration. One of the goals of the National Policy is to establish a harmonized set of greenhouse gas emissions and CAFE standards that automakers can meet with a single national fleet. In addition, EPA has granted the State of California's request for a waiver of pre-emption under the Clean Air Act for its light-duty vehicle greenhouse gas emissions regulatory program.[5]

Past reports in this series have presented fuel economy data only and have not included CO_2 emissions data. With this report, EPA is introducing CO_2 emissions data in anticipation of future CO_2 regulations. Rather than adding CO_2 emissions data to the large number of tables and figures in this report, we are providing a few key summary tables and figures in this section as well as a methodology with which a reader can convert fuel economy values from other sections of this report to equivalent CO_2 emissions levels. EPA also intends to expand its annual Compliance Report to include CO_2 information.[6] Section III and Sections V through VIII of this report, as well as all of the appendices, continue to focus exclusively on fuel economy data.

The light-duty vehicle tailpipe CO_2 emissions data provided in this report represent the sum of three pollutants that EPA and automakers directly measure in the formal emissions certification and fuel economy compliance test programs:

- CO_2 emissions;
- Carbon monoxide emissions, converted to an equivalent CO_2 level on a mass basis by multiplying by a factor of 1.57; and
- Hydrocarbon emissions, converted to an equivalent CO_2 level on a mass basis by multiplying by a factor of approximately 3.17, which is dependent on the measured carbon weight fraction of vehicle test fuel.

---

[2] 549 U.S. 497 (2007).
[3] 74 Federal Register 18886 (April 24, 2009).
[4] 74 Federal Register 49454 (September 28, 2009)
[5] 74 Federal Register 32744 (July 8, 2009).
[6] 2007 Progress Report: Vehicle and Engine Compliance Activities (EPA-420-R-08-11).

EPA-420-R-09-014   November 2009

While including the carbon monoxide and hydrocarbon emissions adds, on average, less than one percent to the tailpipe CO_2-equivalent emissions for late model year light-duty vehicles, they are included in the CO_2 emissions values for three reasons:

- Atmospheric processes convert carbon monoxide and hydrocarbons to CO_2 relatively quickly compared to the much longer atmospheric lifetime of CO_2;
- Carbon monoxide and hydrocarbon emissions are included, along with CO_2, in the "carbon balance" equations that EPA uses to calculate fuel economy values, so they must also be included in the CO_2 values to maintain the mathematical integrity of the equations given below to convert between CO_2 emissions and fuel economy values; and
- Including carbon monoxide and hydrocarbon emissions is consistent with EPA's proposed light-duty vehicle CO_2 emissions standard-setting approach.

EPA routinely collects CO_2, carbon monoxide, and hydrocarbon emissions data as part of its compliance programs. In fact, the individual fuel economy test values that comprise the EPA fuel economy trends database are calculated from a set of "carbon balance" equations based on direct measurement of CO_2, carbon monoxide, and hydrocarbon emissions. Since carbon is neither created nor destroyed in the combustion process, quantifying the various carbon-containing compounds in the vehicle exhaust as well as the carbon weight fraction of the gasoline test fuel allows the precise calculation of the amount of fuel that was combusted in the vehicle engine. Ironically, while the fuel economy values are calculated from CO_2, carbon monoxide, and hydrocarbon emissions data, the historic EPA fuel economy trends database files do not include the direct emissions data. In order to quickly add CO_2 emissions data to this year's report, EPA has back-calculated the CO_2 emissions (and associated carbon monoxide and hydrocarbon emissions, converted to CO_2 on a mass basis) levels from fuel economy values by simply reversing the carbon balance equations. EPA intends to add the direct CO_2 emissions data, for future model years, to the database files for subsequent versions of this report.

As with the fuel economy data in this report, the light-duty vehicle CO_2 emissions values in this report are expressed in two ways: unadjusted/laboratory values (which will be used for CO_2 emissions regulatory compliance under the proposed new standards) and adjusted/real world values (which are used for consumer information and environmental analysis). The CO_2 emissions values do not represent total light-duty vehicle greenhouse gas emissions, as there are other sources of greenhouse gas emissions beyond those included in the tailpipe CO_2 emissions values. It is also important to note that the tailpipe CO_2 emissions data in this report do not reflect greenhouse gas emissions associated with vehicle assembly or component manufacturing, nor upstream fuel-related production or distribution.

The unadjusted/laboratory CO_2 emissions values are the direct emissions data measured over the EPA city and highway tests. The vehicle air conditioner is turned off during these tests. The EPA city and highway tests will likely be used for compliance with future EPA light-duty vehicle CO_2 emissions standards (compliance flexibilities associated with future CO_2 standards will likely allow the use of air conditioning and other credits so that the unadjusted CO_2 emissions data in this report may not align perfectly with the EPA CO_2 standards). For late model year vehicles, the unadjusted CO_2 emissions values represent about 90 percent of total unadjusted light-duty vehicle greenhouse gas emissions. The remaining 10 percent of total light-duty vehicle greenhouse gas emissions is comprised of air conditioner efficiency-related CO_2 emissions (about 4 percent), air conditioner hydrofluorocarbon refrigerant emissions leaks (approximately 4 percent), tailpipe nitrous oxide emissions (about 1 percent), and tailpipe methane emissions (methane is one hydrocarbon compound with a longer atmospheric lifetime and higher global warming potency, but its mass emissions are so low from gasoline vehicles that its potency-adjusted CO_2-equivalent emissions are less than 0.1 percent of total light-duty vehicle greenhouse gas emissions).

EPA-420-R-09-014

The adjusted CO_2 emissions values are calculated by adjusting the direct CO_2 unadjusted/laboratory emissions test data upward to account for the many variables that can affect real world CO_2 emissions. For a detailed discussion of the methodology that EPA uses to convert unadjusted fuel economy values to adjusted fuel economy values, see Appendix A. This same methodology is used to calculate adjusted CO_2 emissions values as well. On average, based on the current fleet mix, adjusted CO_2 emissions levels are about 25 percent higher than unadjusted CO_2 values. Because the adjusted CO_2 values take the impact of air conditioner operation on vehicle tailpipe CO_2 emissions into account, these values represent about 95 percent of total adjusted real world light-duty vehicle greenhouse gas emissions, with the remainder comprised of air conditioner hydrofluorocarbon refrigerant emissions leaks, tailpipe nitrous oxide emissions, and the higher global warming potency associated with tailpipe methane emissions.

Table 3 gives key light-duty vehicle CO_2 emissions data for the entire data series from 1975 through 2009 for cars only, trucks only, and cars and trucks combined. Table 3 is very similar to Table 1, except that the fuel economy data in Table 1 is replaced with CO_2 emissions data in Table 3. Projected industry-wide MY2009 production volumes, which represent the sum of manufacturer-specific production projections provided by automakers to EPA in the spring and summer of 2008, are not shown in Table 3 as it is expected that actual MY2009 production will be 30 to 40 percent lower than projected values due to the recent economic downturn.

EPA-420-R-09-014   19   November 2009

Table 3

Carbon Dioxide Emissions Characteristics of 1975 to 2009 Light Duty Vehicles

Cars

MODEL YEAR	PROD (000)	FRAC	<-Carbon Dioxide Emissions in g/mi->						CO_2/ TON	CO_2/ CU-FT	CO_2/ TON-CU-FT
			LAB CITY	LAB HWY	LAB 55/45	ADJ CITY	ADJ HWY	ADJ COMP			
1975	8237	.806	649	456	563	723	585	658	325		
1976	9722	.788	585	417	508	649	536	597	294		
1977	11300	.800	556	399	486	618	511	570	289	5.2	2.6
1978	11175	.773	517	363	447	574	466	527	294	4.8	2.7
1979	10794	.778	504	362	439	561	464	518	297	4.8	2.7
1980	9443	.835	441	308	381	489	396	447	288	4.3	2.8
1981	8733	.827	413	288	357	457	370	419	272	4.0	2.6
1982	7819	.803	401	274	344	445	351	403	264	3.8	2.5
1983	8002	.777	403	273	344	448	350	403	259	3.7	2.4
1984	10675	.761	398	268	339	441	343	398	257	3.7	2.4
1985	10791	.746	387	259	330	430	332	387	250	3.6	2.3
1986	11015	.717	375	250	319	419	322	375	247	3.5	2.3
1987	10731	.722	372	248	316	419	321	374	246	3.5	2.3
1988	10736	.702	367	243	311	415	315	369	242	3.4	2.3
1989	10018	.693	373	245	316	425	319	375	242	3.5	2.2
1990	8810	.698	380	247	320	434	323	381	240	3.6	2.2
1991	8524	.678	377	245	317	434	322	380	241	3.5	2.3
1992	8108	.666	385	245	322	444	323	385	238	3.6	2.2
1993	8456	.640	377	241	315	438	319	378	236	3.5	2.2
1994	8415	.596	380	241	317	444	321	381	235	3.5	2.2
1995	9396	.620	377	236	314	444	316	380	233	3.5	2.1
1996	7890	.600	378	236	314	449	317	381	232	3.5	2.1
1997	8335	.576	375	236	313	449	317	380	232	3.5	2.1
1998	7972	.551	375	235	312	451	317	380	230	3.5	2.1
1999	8379	.551	380	238	315	458	323	387	230	3.5	2.1
2000	9128	.551	378	238	315	461	326	388	230	3.5	2.1
2001	8408	.539	375	236	313	458	326	387	229	3.5	2.1
2002	8304	.515	371	236	311	458	327	385	227	3.5	2.1
2003	7951	.504	367	233	308	456	323	383	224	3.5	2.1
2004	7538	.480	369	233	308	461	324	385	222	3.5	2.0
2005	8027	.505	360	230	301	454	322	378	219	3.4	2.0
2006	7993	.529	365	231	305	458	323	382	216	3.4	1.9
2007	8085	.529	350	224	293	442	314	369	210	3.4	1.9
2008	7345	.528	347	222	291	438	312	366	207	3.3	1.9
2009	----	.513	344	220	288	434	309	363	206	3.3	1.9

EPA-420-R-09-014

Table 3   (Continued)

Carbon Dioxide Emissions Characteristics of 1975 to 2009 Light Duty Vehicles

Trucks

MODEL YEAR	PROD (000)	FRAC	LAB CITY	LAB HWY	LAB 55/45	ADJ CITY	ADJ HWY	ADJ COMP	CO$_2$/ TON
1975	1987	.194	734	549	649	815	700	766	376
1976	2612	.212	694	526	617	773	673	728	351
1977	2823	.200	635	491	570	705	630	668	323
1978	3273	.227	645	508	585	718	649	690	332
1979	3088	.222	665	530	606	736	680	713	335
1980	1863	.165	541	408	480	604	522	565	292
1981	1821	.173	503	375	446	560	482	524	275
1982	1914	.197	498	369	439	553	474	518	272
1983	2300	.223	489	355	428	542	457	503	267
1984	3345	.239	498	360	435	554	462	512	271
1985	3669	.254	495	357	432	549	459	509	268
1986	4350	.283	473	343	416	530	440	489	262
1987	4134	.278	473	336	412	529	434	486	262
1988	4559	.298	486	339	419	549	440	497	259
1989	4435	.307	491	345	425	559	449	505	258
1990	3805	.302	499	343	429	570	449	511	255
1991	4049	.322	486	334	417	559	438	499	253
1992	4064	.334	499	339	427	573	447	511	252
1993	4754	.360	496	335	423	573	442	508	249
1994	5710	.404	499	340	427	581	451	517	251
1995	5749	.380	508	343	434	592	456	523	250
1996	5254	.400	502	335	427	589	447	517	245
1997	6124	.424	505	340	431	600	456	523	241
1998	6485	.449	502	334	425	596	449	520	243
1999	6839	.449	511	342	434	609	463	532	241
2000	7447	.449	502	339	427	605	458	526	240
2001	7202	.461	505	342	431	609	465	532	238
2002	7815	.485	505	342	431	617	465	532	234
2003	7824	.496	499	335	425	609	460	526	229
2004	8173	.520	502	335	427	621	463	532	226
2005	7866	.495	488	324	415	609	449	517	221
2006	7111	.471	480	320	408	597	442	508	218
2007	7192	.471	475	314	402	589	436	502	211
2008	6554	.472	463	305	392	574	423	488	207
2009	----	.487	458	300	388	570	415	483	205

EPA-420-R-09-014   21   November 2009

Table 3   (Continued)

Carbon Dioxide Emissions Characteristics of 1975 to 2009 Light Duty Vehicles

Cars and Trucks

| MODEL YEAR | PROD (000) | FRAC | <-Carbon Dioxide Emissions in g/mi-> | | | | | | CO_2/ TON |
			LAB CITY	LAB HWY	LAB 55/45	ADJ CITY	ADJ HWY	ADJ COMP	
1975	10224	1.000	665	474	579	741	607	679	335
1976	12334	1.000	608	440	531	675	565	625	306
1977	14123	1.000	572	417	503	635	535	590	296
1978	14448	1.000	546	396	479	607	508	564	302
1979	13882	1.000	539	400	476	600	512	561	306
1980	11306	1.000	457	325	397	508	417	467	289
1981	10554	1.000	429	303	372	475	390	437	273
1982	9732	1.000	420	292	363	466	375	426	266
1983	10302	1.000	422	291	363	469	373	426	261
1984	14020	1.000	422	290	362	468	371	425	260
1985	14460	1.000	414	284	356	460	364	418	255
1986	15365	1.000	403	277	346	451	356	407	251
1987	14865	1.000	400	272	343	450	352	405	251
1988	15295	1.000	403	272	343	455	352	407	247
1989	14453	1.000	410	275	350	466	359	415	247
1990	12615	1.000	416	276	353	475	361	421	245
1991	12573	1.000	412	274	350	474	359	418	245
1992	12172	1.000	423	276	357	488	364	427	242
1993	13211	1.000	420	275	354	487	363	425	241
1994	14125	1.000	428	281	362	499	373	436	241
1995	15145	1.000	426	277	359	501	369	434	239
1996	13144	1.000	428	276	359	505	369	436	237
1997	14459	1.000	430	280	363	513	376	440	236
1998	14458	1.000	432	279	363	516	376	443	236
1999	15218	1.000	439	284	368	526	386	452	235
2000	16574	1.000	434	284	366	525	385	450	235
2001	15610	1.000	435	285	368	528	390	454	233
2002	16119	1.000	436	288	369	535	394	456	230
2003	15775	1.000	433	284	366	532	391	454	227
2004	15711	1.000	438	286	370	544	396	461	224
2005	15893	1.000	424	277	358	530	385	447	220
2006	15105	1.000	419	273	353	524	379	441	217
2007	15277	1.000	409	266	345	511	371	432	211
2008	13900	1.000	402	262	339	502	364	424	207
2009	----	1.000	399	259	337	500	361	422	205

EPA-420-R-09-014

Figure 10 plots the adjusted CO_2 emissions values over time, for cars only, trucks only, and both cars and trucks combined.

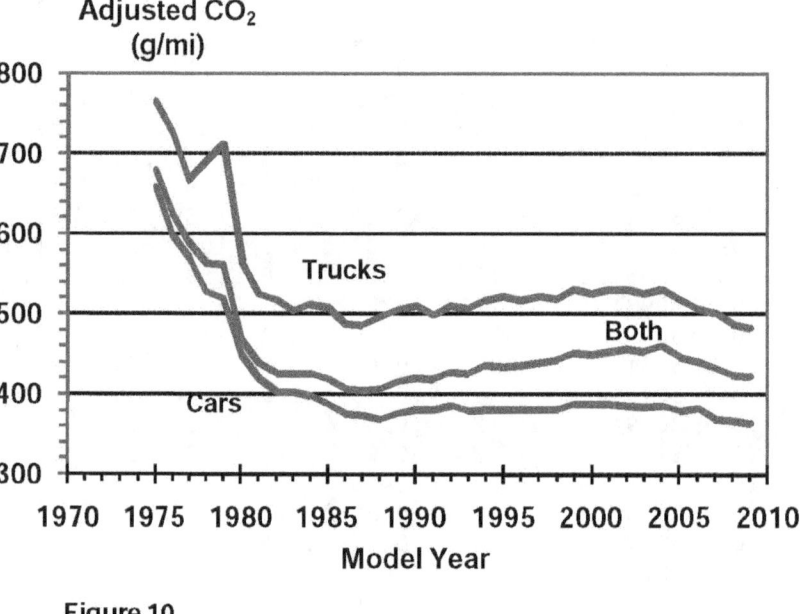

Adjusted CO_2 Emissions by Model Year
(grams/mile)

Figure 10

Table 3 and Figure 10 show that, over the last 35 years, adjusted (real world) CO_2 emissions rates have gone through four distinct phases. Most dramatically, adjusted composite (city/highway) CO_2 emissions rates for the combined car/truck fleet fell sharply from 679 g/mi in MY1975 to 437 g/mi in MY1981, for a 36 percent reduction over 6 years. Adjusted CO_2 emissions continued to decline, though much more slowly, reaching an all-time low of 405 g/mi in MY1987, which represents a 40 percent reduction from MY1975. The trend then reversed, as adjusted CO_2 levels rose slowly over the next 17 years, reaching 461 g/mi in MY2004, a 14 percent increase relative to the MY1987 low. Adjusted CO_2 emissions have decreased for each of the last 5 years. The MY2008 value, based nearly exclusively on final CAFE reports, is 424 g/mi. The preliminary MY2009 value, based on automaker production projections made prior to the beginning of the model year, is 422 g/mi. The preliminary MY2009 value represents an 8 percent reduction relative to MY2004.

Laboratory CO_2 emissions values are also given in Table 3. Because laboratory values do not reflect the changes that EPA made to its methodology for adjusting fuel economy and CO_2 emissions levels for real world estimates for consumers, they are the best metric for evaluating CO_2 emissions trends solely on vehicle design considerations. Based on the 55/45 (city/highway) laboratory CO_2 values in Table 3, the 339 g/mi value in MY2008 and the preliminary MY2009 value of 337 g/mi represent all-time lows.

Table 4 shows key light-duty vehicle characteristics, along with the adjusted composite CO_2 emissions values, for the 1975 through 2009 timeframe for cars only, trucks only, and cars and trucks combined. Table 4 is very similar to Table 2, except that the fuel economy data in Table 2 is replaced with CO_2 emissions data in Table 4.

EPA-420-R-09-014   November 2009

Table 4

Vehicle Size and Design Characteristics of 1975 to 2009 Cars

<---------- Vehicle Characteristics: ----------> <- Percent By: ->

MODEL YEAR	PROD FRAC	ADJ COMP CO_2	VOL CU-FT	WGHT LB	FOOT PRNT SQFT	ENG HP	HP/ WT	0-60 TIME	TOP SPD	VEHICLE SIZE SMALL	MID	LARGE
1975	.806	658		4058		136	.0331	14.2	111	55.4	23.3	21.3
1976	.788	597		4059		134	.0324	14.4	110	55.4	25.2	19.4
1977	.800	570	110	3944		133	.0335	14.0	111	51.9	24.5	23.5
1978	.773	527	109	3588		124	.0342	13.7	111	44.7	34.4	21.0
1979	.778	518	109	3485		119	.0338	13.8	110	43.7	34.2	22.1
1980	.835	447	104	3101		100	.0322	14.3	107	54.4	34.4	11.3
1981	.827	419	106	3076		99	.0320	14.4	106	51.5	36.4	12.2
1982	.803	403	106	3054		99	.0320	14.4	106	56.5	31.0	12.5
1983	.777	403	109	3112		104	.0330	14.0	108	53.1	31.8	15.1
1984	.761	398	108	3099		106	.0339	13.8	109	57.4	29.4	13.2
1985	.746	387	108	3093		111	.0355	13.3	111	55.7	28.9	15.4
1986	.717	375	107	3041		111	.0360	13.2	111	59.5	27.9	12.6
1987	.722	374	107	3031		112	.0365	13.0	112	63.5	24.3	12.2
1988	.702	369	107	3047		116	.0375	12.8	113	64.8	22.3	12.8
1989	.693	375	108	3099		121	.0387	12.5	115	58.3	28.2	13.5
1990	.698	381	107	3176		129	.0401	12.1	117	58.6	28.7	12.8
1991	.678	380	107	3154		132	.0413	11.8	118	61.5	26.2	12.3
1992	.666	385	108	3240		141	.0428	11.5	120	56.5	27.8	15.6
1993	.640	378	108	3207		138	.0425	11.6	120	57.2	29.5	13.3
1994	.596	381	108	3250		143	.0432	11.4	121	58.5	26.1	15.4
1995	.620	380	109	3263		152	.0460	10.9	125	57.3	28.6	14.0
1996	.600	381	109	3282		154	.0464	10.8	125	54.3	32.0	13.6
1997	.576	380	109	3274		156	.0469	10.7	126	55.1	30.6	14.3
1998	.551	380	109	3306		159	.0475	10.6	127	49.4	39.1	11.4
1999	.551	387	109	3365		164	.0481	10.5	128	47.7	39.7	12.6
2000	.551	388	110	3369		168	.0492	10.4	129	47.5	34.3	18.2
2001	.539	387	109	3380		168	.0492	10.3	129	50.9	32.3	16.8
2002	.515	385	109	3391		173	.0504	10.2	131	48.6	36.3	15.1
2003	.504	383	109	3421		176	.0510	10.0	132	50.8	33.4	15.9
2004	.480	385	110	3462		182	.0521	9.8	133	47.4	35.5	17.0
2005	.505	378	111	3463		182	.0518	9.8	133	44.2	38.9	16.8
2006	.529	382	112	3534		194	.0540	9.6	136	46.2	32.9	20.9
2007	.529	369	110	3507		189	.0531	9.6	135	44.6	40.0	15.4
2008	.528	366	110	3527	45.4	193	.0536	9.6	136	44.6	35.9	19.5
2009	.513	363	111	3533		198	.0548	9.5	137	43.8	33.3	23.0

EPA-420-R-09-014

Table 4 (Continued)

Vehicle Size and Design Characteristics of 1975 to 2009 Trucks

					<-------- Vehicle Characteristics: -------->					<----------- Percent By: ----------->					
MODEL YEAR	PROD FRAC	ADJ COMP CO_2	WGHT LB	FOOT PRNT SQFT	ENG HP	HP/ WT	0-60 TIME	TOP SPD	VEHICLE SIZE SMALL	MID	LARGE	VEHICLE TYPE VAN	SUV	PICKUP
1975	.194	766	4072		142	.0349	13.6	114	10.9	24.2	64.9	23.0	9.4	67.6
1976	.212	728	4155		141	.0340	13.8	113	9.0	20.3	70.7	19.2	9.3	71.4
1977	.200	668	4135		147	.0356	13.3	115	11.0	20.4	68.5	18.2	10.0	71.8
1978	.227	690	4151		146	.0351	13.4	114	10.9	22.7	66.3	19.1	11.6	69.3
1979	.222	713	4252		138	.0325	14.3	111	15.2	19.5	65.3	15.6	13.0	71.5
1980	.165	565	3869		121	.0313	14.5	108	28.4	17.6	54.0	13.0	9.9	77.1
1981	.173	524	3806		119	.0311	14.6	108	23.2	19.1	57.7	13.5	7.5	79.1
1982	.197	518	3806		120	.0317	14.5	109	21.1	31.0	47.9	16.2	8.5	75.3
1983	.223	503	3763		118	.0313	14.5	108	16.6	45.9	37.6	16.6	12.6	70.8
1984	.239	512	3782		118	.0310	14.7	108	19.5	46.4	34.1	20.2	18.7	61.1
1985	.254	509	3795		124	.0326	14.1	110	19.2	48.5	32.3	23.3	20.0	56.6
1986	.283	489	3738		123	.0330	14.0	110	23.5	48.5	28.0	24.0	17.8	58.2
1987	.278	486	3713		131	.0351	13.3	113	19.9	59.6	20.6	26.9	21.1	51.9
1988	.298	497	3841		141	.0366	12.9	115	15.0	57.2	27.8	24.8	21.2	53.9
1989	.307	505	3921		146	.0372	12.8	116	13.9	58.9	27.2	28.8	20.9	50.3
1990	.302	511	4005		151	.0377	12.6	117	13.4	57.1	29.6	33.2	18.6	48.2
1991	.322	499	3948		150	.0379	12.6	117	11.4	67.2	21.4	25.5	27.0	47.4
1992	.334	511	4056		155	.0382	12.5	118	10.4	64.0	25.6	30.0	24.7	45.3
1993	.360	508	4073		162	.0398	12.1	120	8.8	65.3	25.9	30.3	27.6	42.1
1994	.404	517	4125		166	.0403	12.0	121	9.8	63.1	27.2	24.8	28.4	46.7
1995	.380	523	4184		168	.0401	12.0	121	8.6	63.5	27.9	28.9	31.6	39.5
1996	.400	517	4225		179	.0423	11.5	124	6.5	67.1	26.4	26.8	36.0	37.2
1997	.424	523	4344		187	.0429	11.4	126	10.1	52.5	37.3	20.7	40.0	39.3
1998	.449	520	4283		187	.0435	11.2	126	8.9	58.7	32.4	23.0	39.8	37.2
1999	.449	532	4412		197	.0446	11.0	128	7.7	55.8	36.5	21.4	41.4	37.2
2000	.449	526	4375		197	.0448	11.0	128	6.7	55.7	37.5	22.7	42.2	35.1
2001	.461	532	4463		209	.0466	10.6	131	6.6	47.6	45.9	17.1	47.9	35.0
2002	.485	532	4546		219	.0482	10.4	134	7.1	43.5	49.4	15.9	53.6	30.5
2003	.496	526	4586		221	.0481	10.4	134	5.9	47.8	46.3	15.7	52.6	31.6
2004	.520	532	4710		236	.0501	10.0	137	5.1	46.2	48.7	11.7	57.7	30.7
2005	.495	517	4668		237	.0505	10.0	137	2.8	47.3	49.9	18.8	51.9	29.2
2006	.471	508	4665		235	.0502	10.0	137	2.0	49.0	49.0	16.4	52.8	30.8
2007	.471	502	4752		248	.0520	9.8	140	2.0	44.9	53.1	11.8	58.8	29.4
2008	.472	488	4710	52.9	247	.0522	9.7	140	2.3	49.7	48.0	11.8	60.8	27.4
2009	.487	483	4712		253	.0534	9.6	142	2.5	44.7	52.9	9.3	65.8	24.9

EPA-420-R-09-014

Table 4   (Continued)

Vehicle Size and Design Characteristics of 1975 to 2009 Light Duty Vehicles

<-------- Vehicle Characteristics: -------->

MODEL YEAR	PROD FRAC	ADJ COMP CO_2	WGHT LB	FOOT PRNT SQFT	ENG HP	HP/ WT	0-60 TIME	TOP SPD
1975	1.000	679	4060		137	.0335	14.1	112
1976	1.000	625	4079		135	.0328	14.3	111
1977	1.000	590	3982		136	.0339	13.8	112
1978	1.000	564	3715		129	.0344	13.6	112
1979	1.000	561	3655		124	.0335	13.9	110
1980	1.000	467	3228		104	.0320	14.3	107
1981	1.000	437	3202		102	.0318	14.4	107
1982	1.000	426	3202		103	.0320	14.4	107
1983	1.000	426	3257		107	.0327	14.1	108
1984	1.000	425	3262		109	.0332	14.0	109
1985	1.000	418	3271		114	.0347	13.5	110
1986	1.000	407	3238		114	.0351	13.4	111
1987	1.000	405	3221		118	.0361	13.1	112
1988	1.000	407	3283		123	.0372	12.8	114
1989	1.000	415	3351		129	.0382	12.5	115
1990	1.000	421	3426		135	.0394	12.2	117
1991	1.000	418	3410		138	.0402	12.1	118
1992	1.000	427	3512		145	.0413	11.8	120
1993	1.000	425	3519		147	.0416	11.8	120
1994	1.000	436	3603		152	.0420	11.7	121
1995	1.000	434	3613		158	.0438	11.3	123
1996	1.000	436	3659		164	.0447	11.1	125
1997	1.000	440	3727		169	.0452	11.0	126
1998	1.000	443	3744		171	.0457	10.9	126
1999	1.000	452	3835		179	.0465	10.7	128
2000	1.000	450	3821		181	.0472	10.6	129
2001	1.000	454	3879		187	.0480	10.5	130
2002	1.000	456	3951		195	.0493	10.3	132
2003	1.000	454	3999		199	.0496	10.2	133
2004	1.000	461	4111		211	.0511	9.9	135
2005	1.000	447	4059		209	.0512	9.9	135
2006	1.000	441	4067		213	.0522	9.8	137
2007	1.000	432	4093		217	.0525	9.7	137
2008	1.000	424	4085	49.0	219	.0529	9.7	138
2009	1.000	422	4108		225	.0541	9.5	139

EPA-420-R-09-014

Table 4 shows that average, combined car/truck, weight and horsepower levels declined significantly from MY1975 through MY1981, with weight decreasing by over 850 pounds (21 percent) and power falling by 35 horsepower (26 percent). Average vehicle weight grew slowly in the 1980s, and more rapidly thereafter, and by MY2004 average weight had reached an all-time high of 4111 pounds. It has remained relatively constant since. Average vehicle horsepower has grown steadily since MY1981. The projected MY2009 level of 225 horsepower represents a 64 percent increase over MY1975, and a 121 percent increase relative to MY1981, which was the all-time low for this data series. Table 4 also shows that average MY2008 footprint values were 45.4 square feet for cars, 52.9 square feet for trucks, and 49.0 square feet for cars and trucks combined.

Table 5 gives average CO_2 emissions performance for the nine highest-production volume marketing groups for model years 2008 and 2009 for cars only, trucks only, and cars and trucks combined. As discussed earlier, EPA has high confidence in the MY2008 data as it is based nearly exclusively on actual production as submitted by automakers to EPA in final CAFE reports. EPA has less confidence in the MY2009 data as it is based on automaker projections of production volumes submitted to EPA prior to the start of the 2009 model year. EPA anticipates that this data will change for all manufacturers after the final MY2009 data has been submitted to EPA, and this final data will be included in next year's version of this report.

Table 5

Carbon Dioxide Emissions by Marketing Group for MY 2008 and MY2009
(g/mi)

Marketing Group	<--------- Model Year 2008 --------->			<--------- Model Year 2009 --------->		
	Cars	Trucks	Cars and Trucks	Cars	Trucks	Cars and Trucks
Honda	328	438	372	334	436	376
Hyundai-Kia	335	447	374	346	447	380
Toyota	316	468	389	324	460	383
Volkswagen	385	546	398	376	464	398
Nissan	351	502	406	344	494	411
BMW	406	480	419	395	496	412
General Motors	386	511	452	378	505	447
Ford	397	499	459	391	475	434
Chrysler	400	494	460	406	496	476
All	366	488	424	363	483	422

For MY2008, Honda had the lowest average car/truck CO_2 emissions performance of 372 g/mi, followed closely by Hyundai-Kia with 374 g/mi. Chrysler had the highest average fleet value of 460 g/mi, or 24 percent higher than Honda, followed by Ford with 459 g/mi. For MY2008 cars, Toyota had the lowest and BMW had the highest average CO_2 emissions. Honda had the lowest average CO_2 emissions performance for MY2008 trucks, while Volkswagen had the highest value.

The relative marketing group rankings for the preliminary MY2009 values are generally similar to those for MY2008. The most notable changes are that the preliminary MY2009 fleetwide value for Ford is 25 g/mi lower than in MY2008, and the preliminary MY2009 value for Chrysler is 16 g/mi higher than in MY2008. It will not be possible to confirm these changes until the final MY2009 CAFE reports become available early next year.

While Tables 3, 4, and 5 provide key summary CO_2 emissions data, EPA recognizes that many users will want the CO_2 emissions values equivalent to the fuel economy values in many other tables in this report. Converting fuel economy values from tables in this report to approximate equivalent CO_2 emissions values is fairly straightforward.

If it is known that a fuel economy value in this report is based on a single gasoline vehicle, or a 100 percent gasoline vehicle fleet, one can calculate the precise corresponding CO_2 value by simply dividing 8887 (which is a typical value for the grams of CO_2 per gallon of gasoline test fuel, assuming all the carbon is converted to CO_2) by the fuel economy value in miles per gallon. For example, 8887 divided by a gasoline vehicle fuel economy of 30 mpg would yield an equivalent CO_2 emissions value of 296 grams per mile.

Since gasoline vehicle production has accounted for 99+ percent of all light-duty vehicle production for all model years since 1975 except for the six years from 1979 through 1984, this simple approach yields very accurate results for most model years.

Diesel fuel has 14.5 percent higher carbon content per gallon than gasoline. To calculate a CO_2 equivalent value for a diesel vehicle, one should divide 10,180 by the diesel vehicle fuel economy value. Accordingly, a 30 mpg diesel vehicle would have a CO_2 equivalent value of 339 grams per mile.

Table 6 should be used by those who want to make the most accurate conversions of industry-wide fuel economy values to CO_2 emissions values. Table 6 gives model year-specific industry-wide values for grams of CO_2 per gallon based on actual light-duty gasoline and diesel vehicle production in that year. Using these model year-specific values and dividing by the fuel economy value in miles per gallon will allow accurate conversions of industry-wide fuel economy values to industry-wide CO_2 emissions values.

Readers will have to make judgment calls about how to best convert fuel economy values that do not represent industry-wide values (e.g., just small vehicles or vehicles with 5-speed automatic transmissions). If the user knows the gasoline/diesel production volume fractions of the individual database component, it is best to generate a weighted value of grams of CO_2 per gallon based on the 8887 (gasoline) and 10,180 (diesel) factors discussed above. Otherwise, the reader can choose between the model year-specific weighting in Table 6 (which implicitly assumes that the diesel fraction in the database component of interest is similar to that for the overall fleet in that year) or the gasoline value of 8887 (implicitly assuming no diesels in that database component). In nearly all cases, any error associated with either of these approaches will be relatively small.

EPA-420-R-09-014

Table 6

Factors for Converting Industry-wide Fuel Economy Values from this Report to Carbon Dioxide Emissions Values

Model Year	Gasoline Market Share (Percentage)	Diesel Market Share (Percentage)	Weighted CO_2 per Gallon (grams)
1975	99.8	.2	8890
1976	99.8	.2	8890
1977	99.6	.4	8892
1978	99.1	.9	8899
1979	98.0	2.0	8913
1980	95.7	4.3	8943
1981	94.1	5.9	8963
1982	94.4	5.6	8959
1983	97.3	2.7	8922
1984	98.2	1.8	8910
1985	99.1	.9	8899
1986	99.6	.4	8892
1987	99.7	.3	8891
1988	99.9	.1	8888
1989	99.9	.1	8888
1990	99.9	.1	8888
1991	99.9	.1	8888
1992	99.9	.1	8888
1993	100.0	.0	8887
1994	100.0	.0	8887
1995	100.0	.0	8887
1996	99.9	.1	8888
1997	99.9	.1	8888
1998	99.9	.1	8888
1999	99.9	.1	8888
2000	99.9	.1	8888
2001	99.9	.1	8888
2002	99.8	.2	8890
2003	99.8	.2	8890
2004	99.9	.1	8888
2005	99.7	.3	8891
2006	99.6	.4	8892
2007	99.9	.1	8888
2008	99.9	.1	8888
2009	99.5	.5	8893

EPA-420-R-09-014

V. Fuel Economy Trends by Vehicle Type, Size, and Weight

Table 1 showed that for the past several years trucks have accounted for about 50 percent of the light-duty vehicles produced each year. MY2004 was the peak year for trucks with 52 percent market share, and trucks have been between 47 and 50 percent since. Considering the five classes: cars, wagons, sports utility vehicles (SUVs), vans, and pickups, since 1975 the biggest overall increase in market share has been for SUVs, up from less than two percent in 1975 to over 30 percent based on a 3-year moving average (see Figure 11 and Table 7). The biggest overall decrease has been for cars, down from 71 percent of the fleet in 1975 to 47 percent. By comparison, the production fraction for pickup trucks has remained relatively constant at about 12 percent of the market.

Figures 12 to 16 compare 3-year moving average production fractions by vehicle type and size with the fleet again stratified into five vehicle types: cars (i.e., coupes, sedans, and hatchbacks), station wagons, vans, SUVs, and pickup trucks; and three vehicle sizes: small, midsize, and large. As shown in Figure 12, large cars accounted for about 20 percent of all car production in the late 1970s, but their share of the car market dropped in the early 1980s to about 12 percent of the market where it remained for about two decades, but has since increased back to about 20 percent. Within the car segment, the market share for small cars peaked in the late 1980s at about 65 percent and is now lower than at anytime since 1975.

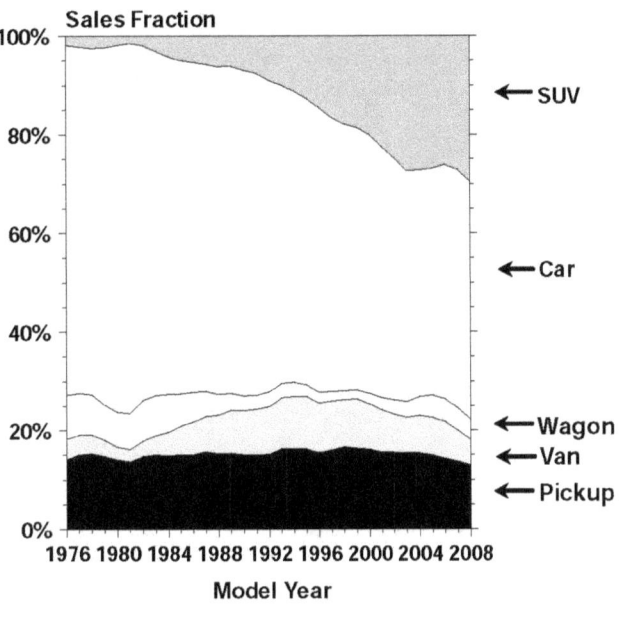

Figure 11

EPA-420-R-09-014

Large wagons accounted for more than 20 percent of the wagon segment of the market in the late 1970s but then lost market share relatively consistently and were not produced at all between 1996 and 2004 when they re-emerged. They now account for about five percent of all wagons, but less than one percent of all light vehicles. Similarly (see Figure 14), large vehicles accounted for nearly 40 percent of all vans through the early 1980s compared to less than 10 percent the past few years. Small vans have never had a significant market share, and none have been produced in recent years. Figures 15 and 16 show that the longer term trend of increased market share for both large SUVs and pickups has levelled off in the last few years.

Table 7 compares the production fractions by vehicle type and size on a different basis, that for the total market. Since 1975, the largest increases in production fractions have been for midsize and large SUVs. These two classes are expected to account for 30 percent of all light vehicles built this year, compared to combined totals of about 1.3 and 4.5 percent in 1975 and 1988, respectively. Conversely, the largest production fraction decrease has occurred for small cars which accounted for 40 percent of all light-duty vehicles produced in 1975 and over 43 percent in 1988, but less than 20 percent this year.

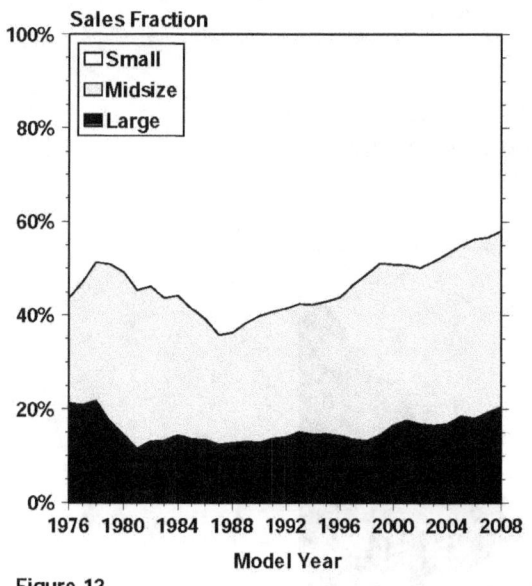

Figure 12

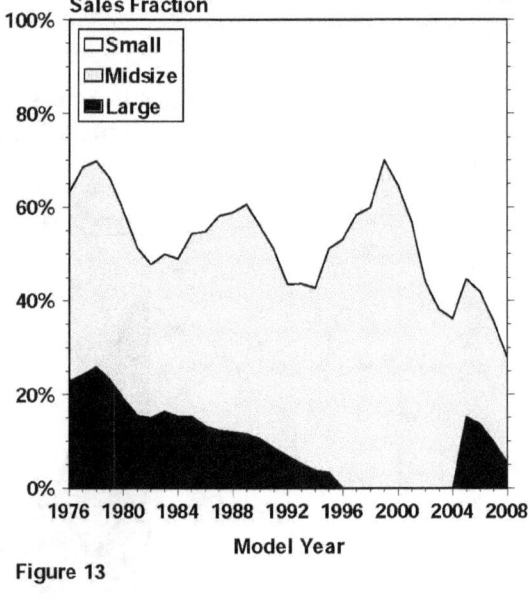

Figure 13

Van Sales Fraction by Vehicle Size
(Three Year Moving Average)

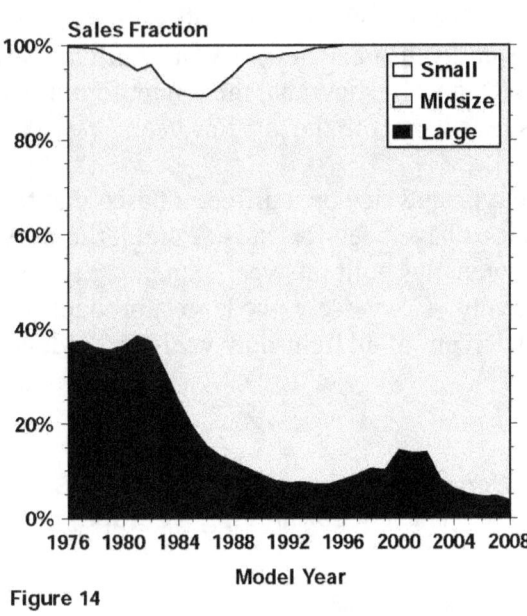

Figure 14

SUV Sales Fraction by Vehicle Size
(Three Year Moving Average)

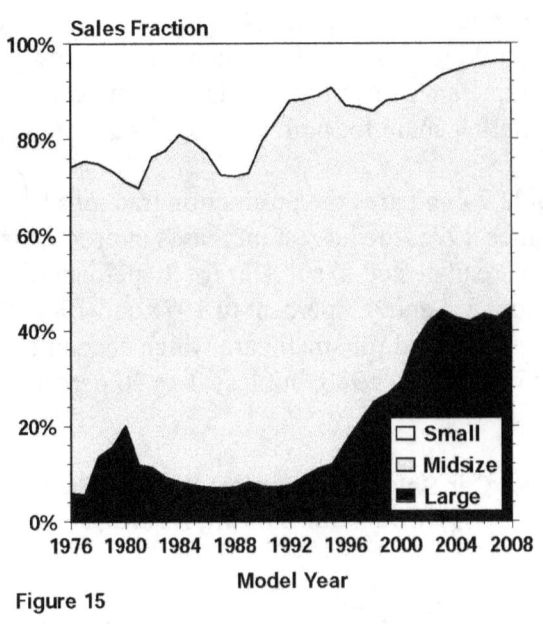

Figure 15

Pickup Sales Fraction by Vehicle Size
(Three Year Moving Average)

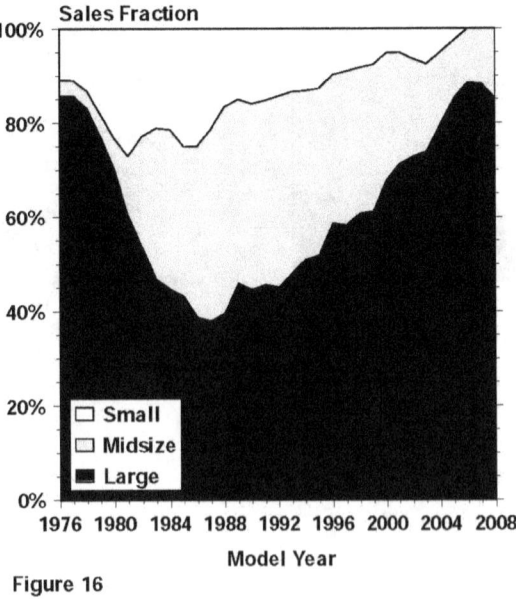

Figure 16

EPA-420-R-09-014   32   November 2009

Table 7

Production Fractions of MY1975, MY1988 and MY2009
Light Duty Vehicles by Vehicle Size and Type

| | | Production Fraction | | | Differences in Production Fraction | | |
Vehicle Type	Size	1975	1988	2009	From 1975 To 2009	From 1975 To 1988	From 1988 To 2009
Car	Small	40.0%	43.8%	19.0%	-21.0%	3.9%	-24.8%
	Midsize	16.0%	13.8%	16.3%	.3%	-2.1%	2.5%
	Large	15.2%	8.5%	11.7%	-3.5%	-6.7%	3.2%
	All	71.1%	66.2%	47.0%	-24.1%	-5.0%	-19.2%
Wagon	Small	4.7%	1.7%	3.4%	-1.2%	-3.0%	1.8%
	Midsize	2.8%	1.9%	.8%	-2.1%	-1.0%	-1.1%
	Large	1.9%	.5%	.0%	-1.9%	-1.4%	-0.4%
	All	9.4%	4.0%	4.2%	-5.2%	-5.4%	.2%
Van	Small	.0%	.4%	.0%	.0%	.3%	-0.4%
	Midsize	3.0%	6.2%	4.4%	1.4%	3.2%	-1.8%
	Large	1.5%	.9%	.2%	-1.3%	-0.6%	-0.7%
	All	4.5%	7.4%	4.5%	.1%	2.9%	-2.9%
SUV	Small	.5%	1.9%	1.2%	.7%	1.4%	-0.7%
	Midsize	1.2%	4.0%	15.4%	14.2%	2.8%	11.4%
	Large	.1%	.5%	15.5%	15.4%	.3%	15.0%
	All	1.8%	6.3%	32.1%	30.2%	4.5%	25.7%
Pickup	Small	1.6%	2.2%	.0%	-1.6%	.7%	-2.2%
	Midsize	.5%	6.9%	2.0%	1.5%	6.3%	-4.9%
	Large	11.0%	7.0%	10.1%	-0.9%	-4.1%	3.1%
	All	13.1%	16.1%	12.1%	-1.0%	2.9%	-3.9%
All	Trucks	19.4%	29.8%	48.7%	29.3%	10.4%	18.9%

Figures 17 through 21 show 3-year moving average trends in performance, weight, and adjusted fuel economy for cars, wagons, vans, SUVs, and pickups. For all five vehicle types, there has been a clear long term trend towards increased weight, moderating since 2005 for wagons and SUVs.

Table 8 shows the lowest, average, and highest adjusted mpg performance by vehicle class and size for three selected years. For both 1988 and 2009, the mpg performance is such that the midsize vehicles in all classes have better fuel economy than the corresponding entry for small vehicles in 1975. In Table 9, the percentage changes obtainable from the entries in Table 8 are presented. Average mpg for four classes (midsize cars, large cars, midsize wagons, and midsize SUVs) have improved over 80 percent since 1975. Since 1988, average fuel economy has decreased for small wagons, large wagons, small SUVs, and midsize pickups, and the largest improvements in average mpg has been over 20 percent for midsize and large SUVs, respectively. Tables 10 and 11 present this same data in terms of fuel consumption.

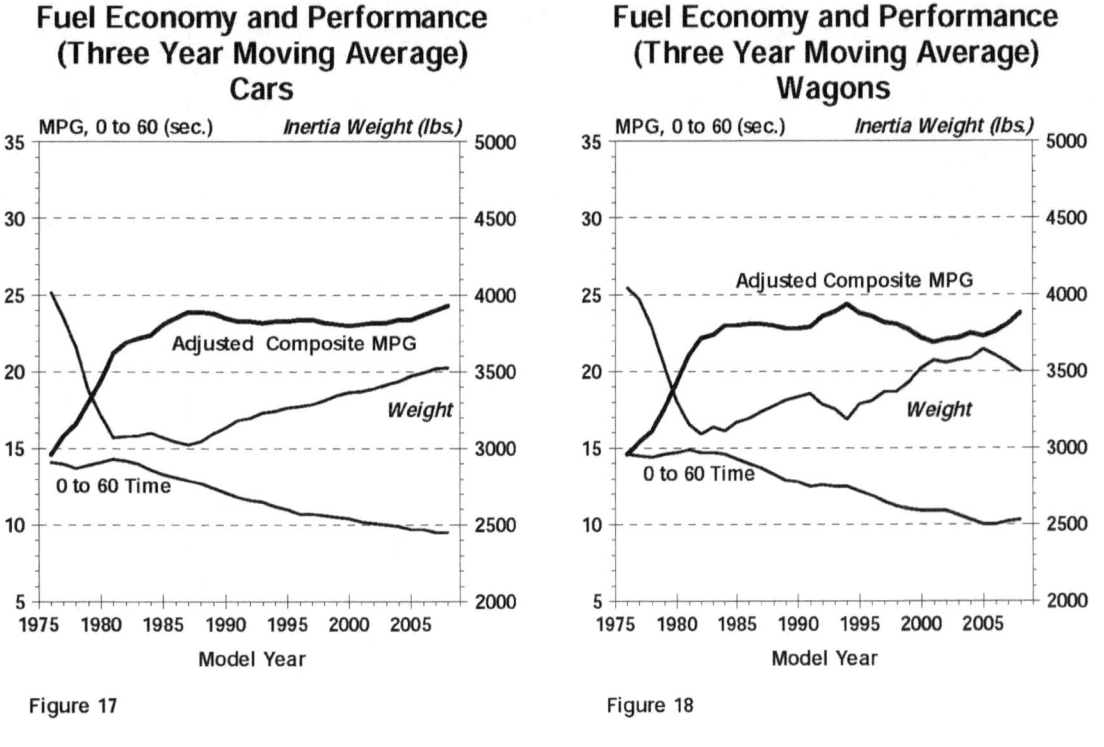

Figure 17

Figure 18

EPA-420-R-09-014

34

November 2009

Fuel Economy and Performance
(Three Year Moving Average)
Vans

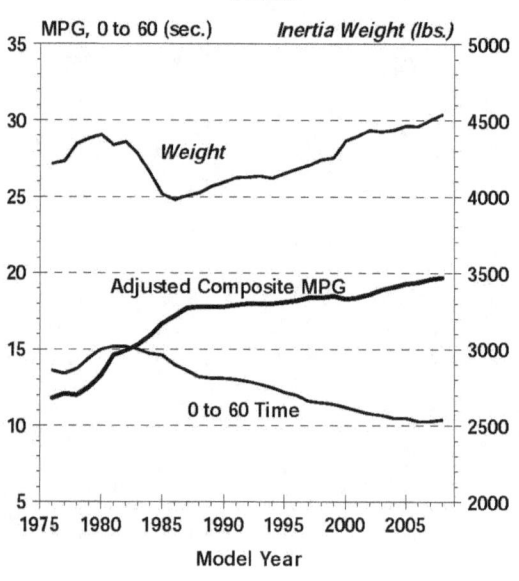

Figure 19

Fuel Economy and Performance
(Three Year Moving Average)
SUVs

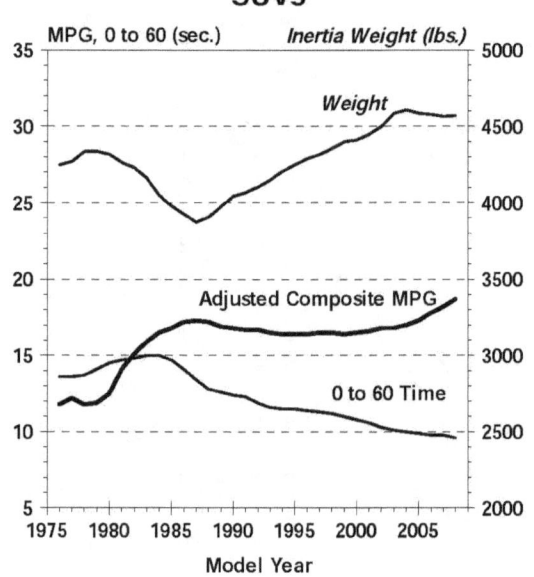

Figure 20

Fuel Economy and Performance
(Three Year Moving Average)
Pickups

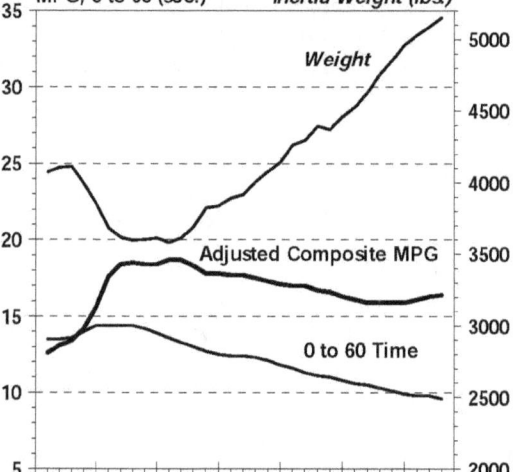

Figure 21

EPA-420-R-09-014   35   November 2009

Table 8

Lowest, Average and Highest Adjusted Fuel Economy by Vehicle Type and Size

Vehicle Type	Size	1975			1988			2009		
		Low.	Avg.	High.	Low.	Avg.	High.	Low.	Avg.	High.
Car	Small	8.6	15.6	28.3	7.5	25.7	54.4	10.4	25.7	42.9
	Midsize	8.6	11.6	18.4	10.5	22.6	27.7	11.9	25.1	46.2
	Large	8.4	11.2	14.6	10.0	20.6	26.0	12.1	22.2	26.4
	All	8.4	13.4	28.3	7.5	24.2	54.4	10.4	24.5	46.2
Wagon	Small	11.8	19.1	24.1	17.1	26.3	33.2	19.2	25.5	35.0
	Midsize	8.4	11.3	25.0	17.5	22.2	27.7	15.4	22.0	24.8
	Large	8.4	10.2	12.8	19.2	19.4	19.4	17.0	17.4	20.9
	All	8.4	13.8	25.0	17.1	23.3	33.2	15.4	24.7	35.0
Van	Small	16.2	17.5	18.5	15.5	20.6	25.0			
	Midsize	8.2	11.3	18.4	11.3	18.4	23.4	18.4	20.1	21.1
	Large	8.9	10.7	14.5	9.9	14.3	16.8	11.2	15.8	17.4
	All	8.2	11.1	18.5	9.9	17.9	25.0	11.2	19.8	21.1
SUV	Small	10.2	13.7	16.3	15.6	20.4	27.7	16.7	18.7	23.1
	Midsize	8.2	10.2	18.4	10.2	16.5	23.6	12.7	20.2	32.0
	Large	7.9	10.3	13.7	12.2	14.0	18.8	10.0	17.9	22.0
	All	7.9	11.0	18.4	10.2	17.2	27.7	10.0	19.0	32.0
Pickup	Small	13.0	19.2	20.8	13.3	21.0	24.6			
	Midsize	17.8	17.9	18.0	15.3	21.3	25.9	15.9	19.5	23.7
	Large	7.6	11.1	18.5	9.8	15.2	21.0	11.5	16.2	21.5
	All	7.6	11.9	20.8	9.8	18.1	25.9	11.5	16.7	23.7
All	Cars	8.4	13.5	28.3	7.5	24.1	54.4	10.4	24.5	46.2
All	Trucks	7.6	11.6	20.8	9.8	17.9	27.7	10.0	18.4	32.0
All	Vehicles	7.6	13.1	28.3	7.5	21.9	54.4	10.0	21.1	46.2

Table 9

Percent Change in Lowest, Average and Highest Adjusted Fuel Economy by Vehicle Type and Size

Vehicle Type	Size	From 1975 to 2009			From 1975 to 1988			From 1988 to 2009		
		Low.	Avg.	High.	Low.	Avg.	High.	Low.	Avg.	High.
Car	Small	21%	65%	52%	-12%	65%	92%	39%	0%	-20%
	Midsize	38%	116%	151%	22%	95%	51%	13%	11%	67%
	Large	44%	98%	81%	19%	84%	78%	21%	8%	2%
	All	24%	83%	63%	-10%	81%	92%	39%	1%	-14%
Wagon	Small	63%	34%	45%	45%	38%	38%	12%	-2%	5%
	Midsize	83%	95%	0%	108%	96%	11%	-11%	0%	-9%
	Large	102%	71%	63%	129%	90%	52%	-10%	-9%	8%
	All	83%	79%	40%	104%	69%	33%	-9%	6%	5%
Van	Small				-3%	18%	35%			
	Midsize	124%	78%	15%	38%	63%	27%	63%	9%	-9%
	Large	26%	48%	20%	11%	34%	16%	13%	10%	4%
	All	37%	78%	14%	21%	61%	35%	13%	11%	-15%
SUV	Small	64%	36%	42%	53%	49%	70%	7%	-7%	-16%
	Midsize	55%	98%	74%	24%	62%	28%	25%	22%	36%
	Large	27%	74%	61%	54%	36%	37%	-17%	28%	17%
	All	27%	73%	74%	29%	56%	51%	-1%	10%	16%
Pickup	Small				2%	9%	18%			
	Midsize	-10%	9%	32%	-13%	19%	44%	4%	-7%	-7%
	Large	51%	46%	16%	29%	37%	14%	17%	7%	2%
	All	51%	40%	14%	29%	52%	25%	17%	-7%	-7%
All	Cars	24%	81%	63%	-10%	79%	92%	39%	2%	-14%
All	Trucks	32%	59%	54%	29%	54%	33%	2%	3%	16%
All	Vehicles	32%	61%	63%	0%	67%	92%	33%	-3%	-14%

EPA-420-R-09-014   37   November 2009

Table 10

Adjusted Fuel Consumption (Gallons/100 miles) by Vehicle Type and Size

Vehicle Type	Size	1975			1988			2009		
		Low.	Avg.	High.	Low.	Avg.	High.	Low.	Avg.	High.
Car	Small	11.6	6.4	3.5	13.3	3.9	1.8	9.6	3.9	2.3
	Midsize	11.6	8.6	5.4	9.5	4.4	3.6	8.4	4.0	2.2
	Large	11.9	8.9	6.8	10.0	4.9	3.8	8.3	4.5	3.8
	All	11.9	7.5	3.5	13.3	4.1	1.8	9.6	4.1	2.2
Wagon	Small	8.5	5.2	4.1	5.8	3.8	3.0	5.2	3.9	2.9
	Midsize	11.9	8.8	4.0	5.7	4.5	3.6	6.5	4.5	4.0
	Large	11.9	9.8	7.8	5.2	5.2	5.2	5.9	5.7	4.8
	All	11.9	7.2	4.0	5.8	4.3	3.0	6.5	4.0	2.9
Van	Small	6.2	5.7	5.4	6.5	4.9	4.0			
	Midsize	12.2	8.8	5.4	8.8	5.4	4.3	5.4	5.0	4.7
	Large	11.2	9.3	6.9	10.1	7.0	6.0	8.9	5.1	4.7
	All	12.2	9.0	5.4	10.1	5.6	4.0	8.9	5.1	4.7
SUV	Small	9.8	7.3	6.1	6.4	4.9	3.6	6.0	5.3	4.3
	Midsize	12.2	9.8	5.4	9.8	6.1	4.2	7.9	5.0	3.1
	Large	12.7	9.7	7.3	8.2	7.1	5.3	10.0	5.6	4.5
	All	12.7	9.1	5.4	9.8	5.8	3.6	10.0	5.3	3.1
Pickup	Small	7.7	5.2	4.8	7.5	4.8	4.1			
	Midsize	5.6	5.6	5.6	6.5	4.7	3.9	6.3	5.1	4.2
	Large	13.2	9.0	5.4	10.2	6.6	4.8	8.7	6.2	4.7
	All	13.2	8.4	4.8	10.2	5.5	3.9	8.7	6.0	4.2
All	Cars	11.9	7.4	3.5	13.3	4.1	1.8	9.6	4.1	2.2
All	Trucks	13.2	8.6	4.8	10.2	5.6	3.6	10.0	5.4	3.1
All	Vehicles	13.2	7.6	3.5	13.3	4.6	1.8	10.0	4.7	2.2

EPA-420-R-09-014

Table 11

Percent Change* in Adjusted Fuel Consumption by Vehicle Type and Size

Vehicle Type	Size	From 1975 to 2009			From 1975 to 1988			From 1988 to 2009		
		Low	Avg.	High	Low	Avg.	High.	Low	Avg.	High
Car	Small	17%	39%	17%	-15%	39%	49%	28%	0%	-28%
	Midsize	28%	53%	59%	18%	49%	33%	12%	9%	39%
	Large	30%	49%	44%	16%	45%	44%	17%	8%	0%
	All	19%	45%	37%	-12%	45%	49%	28%	0%	-22%
Wagon	Small	39%	25%	29%	32%	27%	27%	10%	-3%	3%
	Midsize	45%	49%	0%	52%	49%	10%	-14%	0%	-11%
	Large	50%	42%	38%	56%	47%	33%	-13%	-10%	8%
	All	45%	44%	28%	51%	40%	25%	-12%	7%	3%
Van	Small				-5%	14%	26%			
	Midsize	56%	43%	13%	28%	39%	20%	39%	7%	-9%
	Large	21%	45%	32%	10%	25%	13%	12%	27%	22%
	All	27%	43%	13%	17%	38%	26%	12%	9%	-18%
SUV	Small	39%	27%	30%	35%	33%	41%	6%	-8%	-19%
	Midsize	35%	49%	43%	20%	38%	22%	19%	18%	26%
	Large	21%	42%	38%	35%	27%	27%	-22%	21%	15%
	All	21%	42%	43%	23%	36%	33%	-2%	9%	14%
Pickup	Small				3%	8%	15%			
	Midsize	-13%	9%	25%	-16%	16%	30%	3%	-9%	-8%
	Large	34%	31%	13%	23%	27%	11%	15%	6%	2%
	All	34%	29%	13%	23%	35%	19%	15%	-9%	-8%
All	Cars	19%	45%	37%	-12%	45%	49%	28%	0%	-22%
All	Trucks	24%	37%	35%	23%	35%	25%	2%	4%	14%
All	Vehicles	24%	38%	37%	-1%	39%	49%	25%	-2%	-22%

*Note: A Negative Change indicates that the fuel consumption has increased.

Cars and light trucks with conventional drivetrains have a fuel consumption and weight relationship which is well known and is shown on Figures 22 and 23. Fuel consumption increases linearly with weight. Because vehicles with different propulsion systems, i.e., diesels and hybrids, occupy a different place on such a fuel consumption and weight plot, the data for hybrid and diesel vehicles are plotted separately and excluded from the regression lines shown on the graphs. At constant weight, MY2008 cars consume about 30 to 40 percent less fuel per mile than their MY1975 counterparts.

On this same constant weight basis, this year's cars with diesel engines nominally consume 20 – 25 percent less fuel than the conventionally powered ones, while this year's hybrid cars are about 30 – 40 percent better. Similarly, at constant weight this year's conventionally powered trucks achieve about 40 percent better fuel consumption than MY1975 vehicles did.

Figures 24 and 25 show that the relationship between interior volume and fuel consumption is currently not as important as it used to be. The data points on both of these graphs exclude two seaters and represent production weighted average fuel consumption calculated at increments of 1.0 cu. ft. As was done for Figures 22 and 23, the data points for hybrid and diesel vehicles were plotted separately from those for the conventionally powered vehicles.

As discussed above, EPA is including vehicle footprint data for the first time. We are only reporting MY2008 footprint data in this report. Figures 26 and 27 show laboratory 55/45 fuel consumption versus footprint for cars and trucks, respectively, again with the regression lines excluding the hybrid and diesel data points. Car fuel consumption is more sensitive to footprint than truck fuel consumption. For a given footprint, trucks generally have somewhat higher fuel consumption than cars.

Figures 28 and 29 show the improvement that occurred between 1975 and 2009 for fuel consumption as a function of 0-to-60 time for cars and trucks. Figures 30 and 31 compare Ton-MPG data versus 0-to-60 time and show that at constant vehicle performance, there has been substantial improvement in Ton-MPG, particularly for hybrid and diesel vehicles.

EPA-420-R-09-014

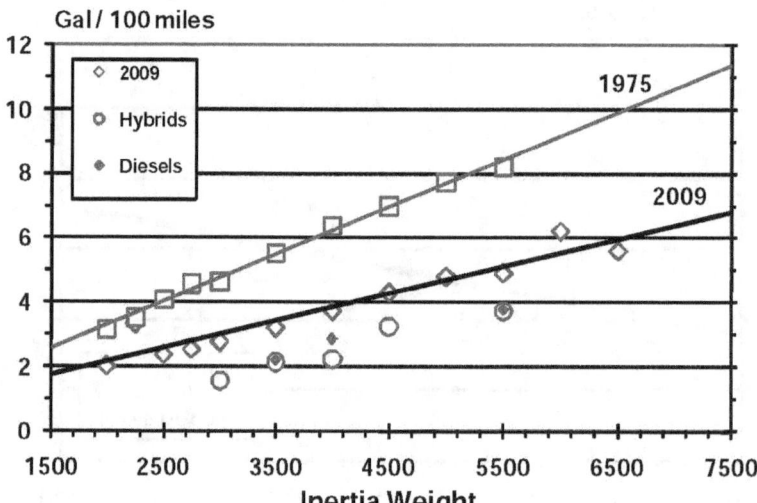

Figure 22

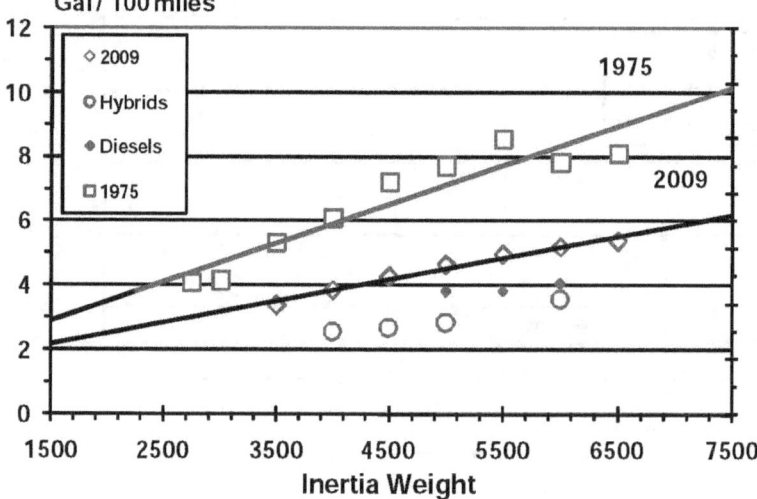

Figure 23

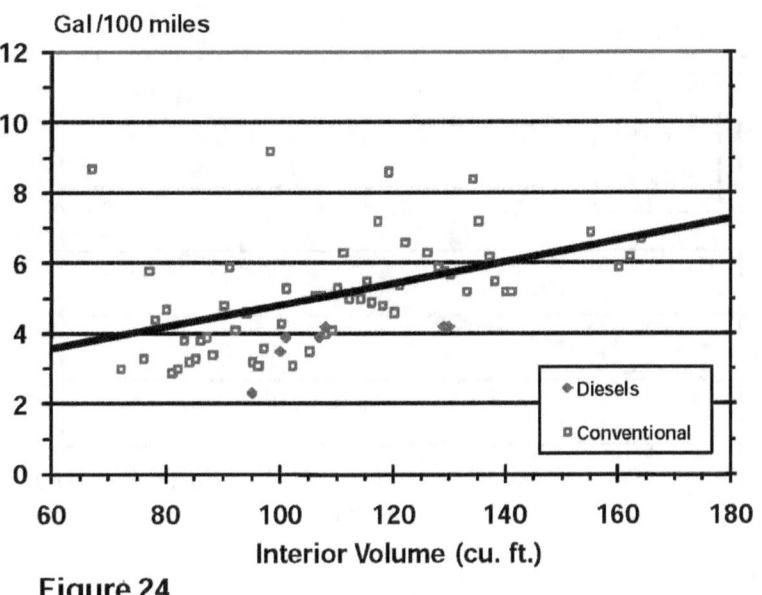

Figure 24

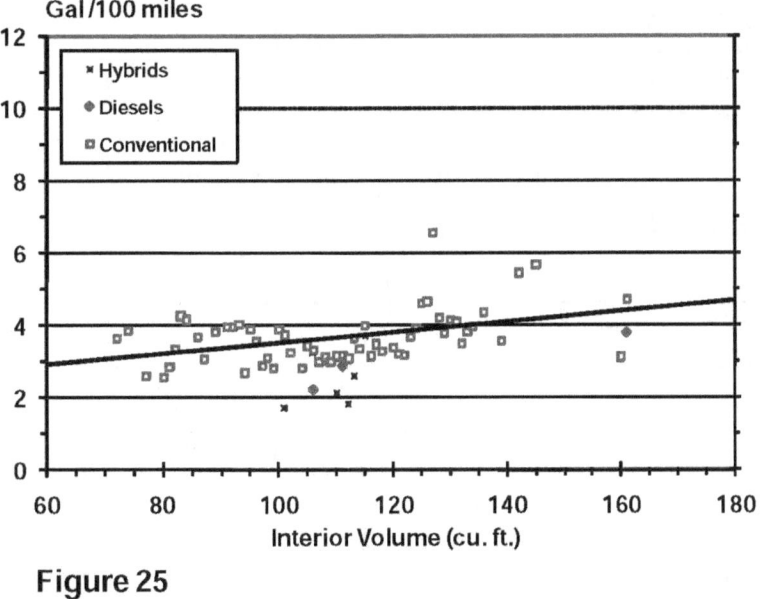

Figure 25

EPA-420-R-09-014   42   November 2009

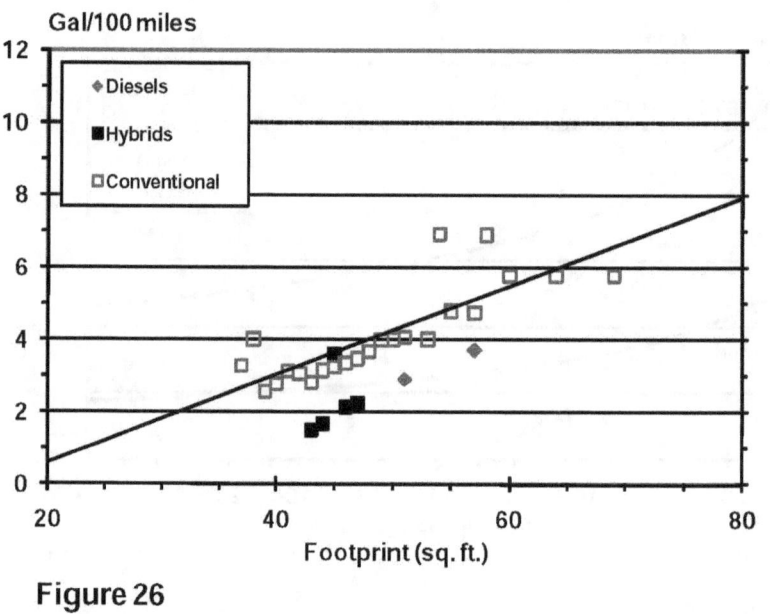

Figure 26

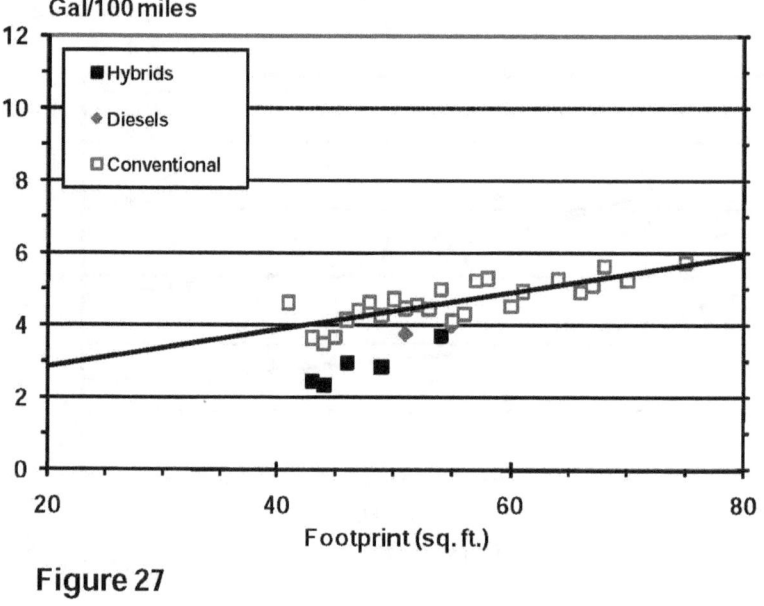

Figure 27

EPA-420-R-09-014   43   November 2009

Laboratory 55/45 Fuel Consumption
vs 0 to 60 Time
MY1975 and MY2009 Cars

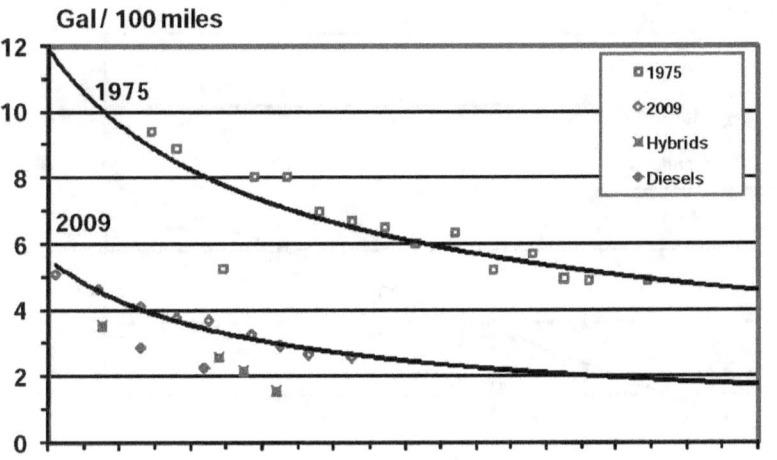

Figure 28

Laboratory 55/45 Fuel Consumption
vs 0 to 60 Time
MY1975 and MY2009 Trucks

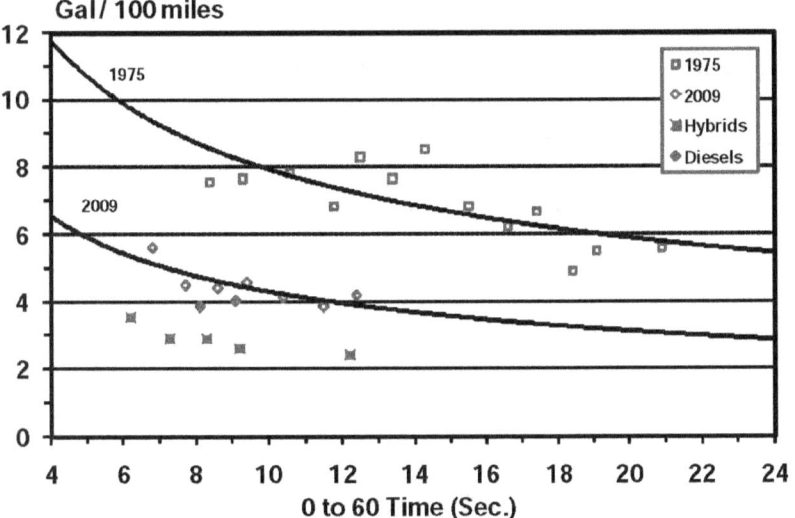

Figure 29

Ton-MPG vs 0 to 60 Time
MY1975 and MY2009 Cars

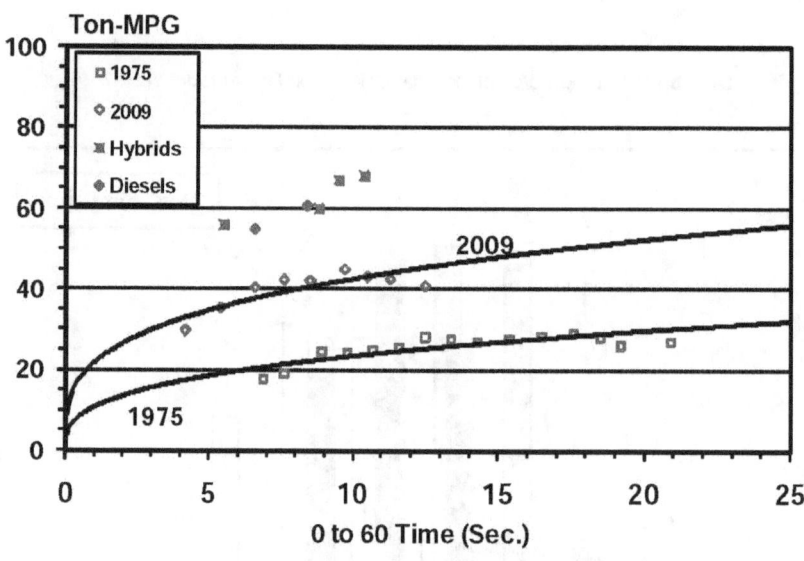

Figure 30

Ton-MPG vs 0 to 60 Time
MY1975 and MY2009 Trucks

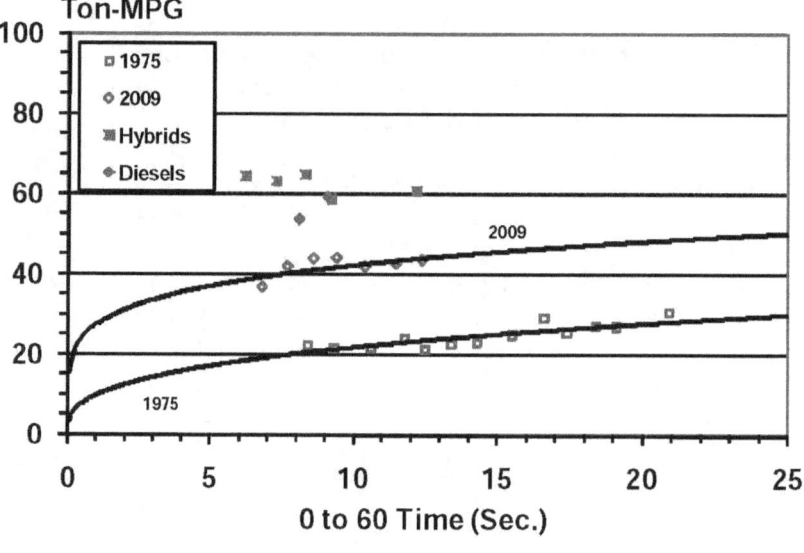

Figure 31

Figure 32 and Table 12 show some of the changes in the distribution of inertia weight that have occurred over the years for the light-duty fleet. In 1975, 13 percent of all light-duty vehicles had inertia weights of less than 3000 lb compared to less than 5 percent in 2009. Since 1988, market share for vehicles with weight of 5000 pounds or more has increased from 3 percent to 20 percent.

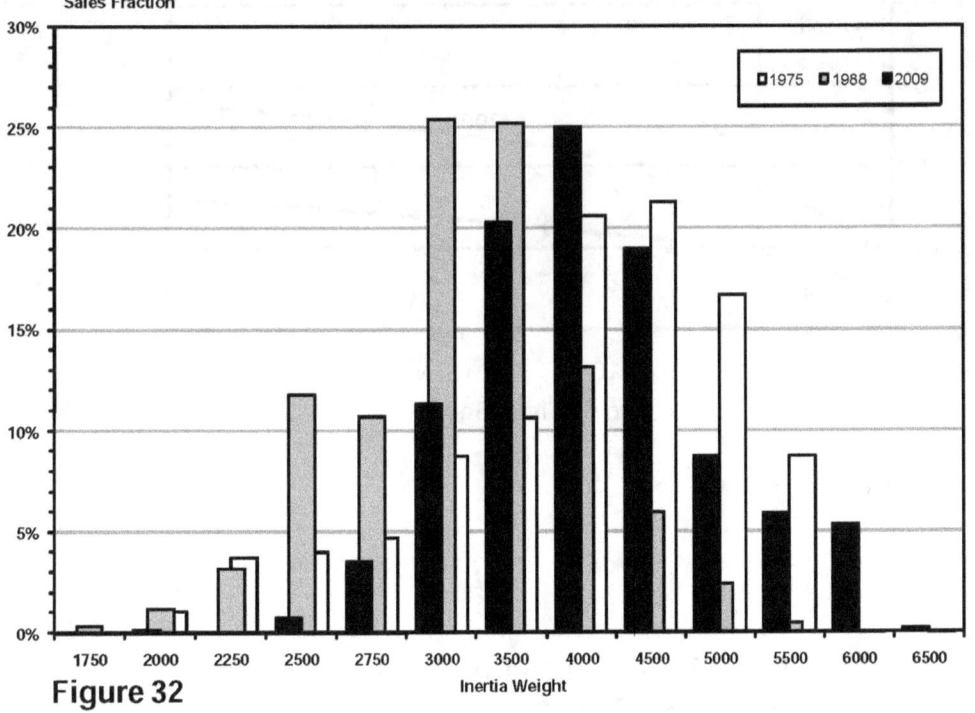

Figure 32

Table 12

Light Vehicle Production Fraction
by Inertia Weight Class
for Three Model Years

Inertia	<----	Model Year	---->
Weight	1975	1988	2009
<3000	13.4%	27.2%	4.4%
3000	8.7%	25.4%	11.3%
3500	10.6%	25.2%	20.3%
4000	20.6%	13.2%	25.0%
4500	21.3%	6.0%	19.0%
5000	16.7%	2.4%	8.7%
5500	8.7%	.5%	6.0%
>5500	.0%	.0%	5.5%
Avg Wt.	4060	3283	4108

EPA-420-R-09-014

Figures 33 through 37 provide an indication of the market share of different weight vehicles within the different classes using 3-year moving averages. Trends within classes are shown which underlie the increasing weight shown by the fleet as a whole. In 1975, about 40 percent of the cars were in inertia weight classes greater than 4000 pounds, compared to less than 5 percent this year. For MY2008, three weight classes (3000, 3500, and 4000 lbs) account for over 90 percent of all cars. Conversely, the market share of trucks in the inertia weight classes of 4500 lb or more have increased substantially, and these vehicles currently account for over 70 percent of all trucks, compared to about 30 percent in 1975. Figures 35, 36, and 37 provide additional details of the truck data presented in Figure 34 for vans, SUVs, and pickups respectively. Appendices D, E, and F contain a series of tables describing light-duty vehicles at the vehicle size/type level of stratification in more detail; Appendix G provides similar data by vehicle type and inertia weight class.

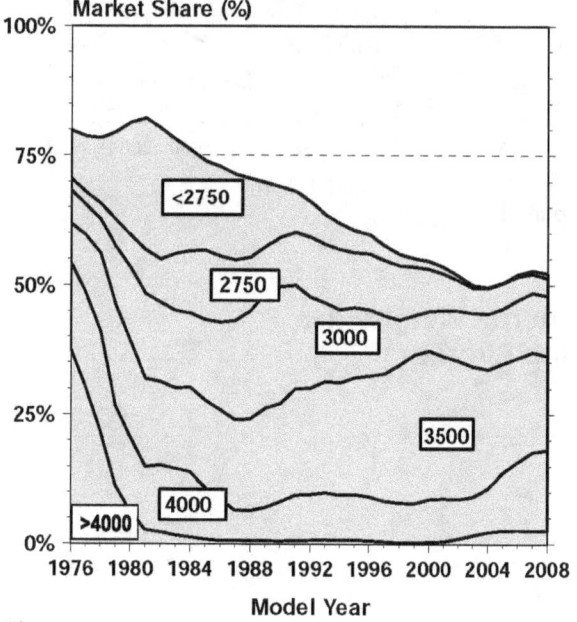

Car Market Share by Inertia Weight Class (Three Year Moving Average)

Figure 33

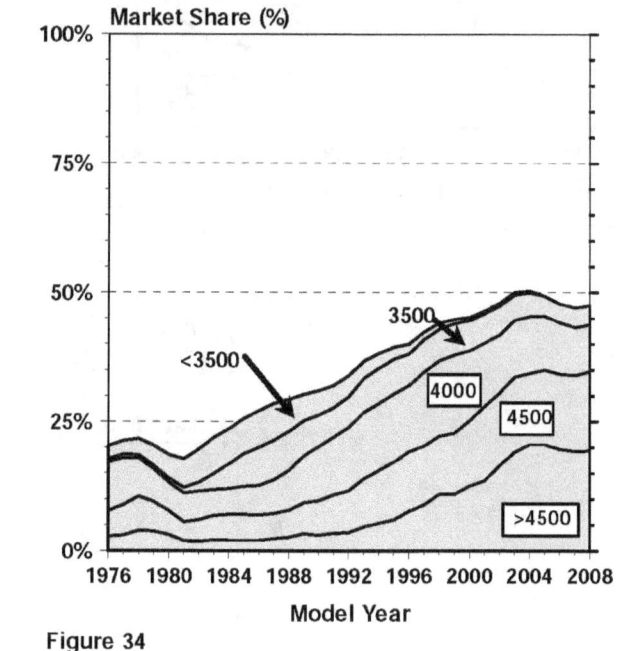

Truck Market Share by Inertia Weight Class (Three Year Moving Average)

Figure 34

EPA-420-R-09-014   47   November 2009

Van Market Share by Inertia Weight Class
(Three Year Moving Average)

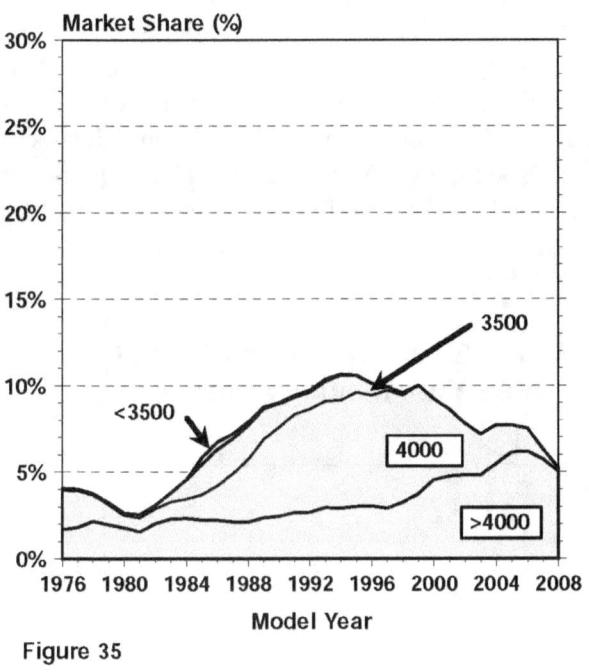

Figure 35

SUV Market Share by Inertia Weight Class
(Three Year Moving Average)

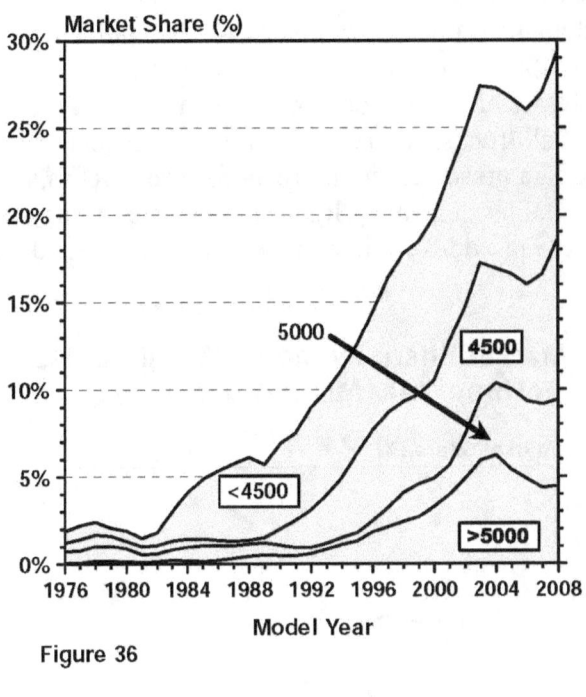

Figure 36

Pickup Market Share by Inertia Weight Class
(Three Year Moving Average)

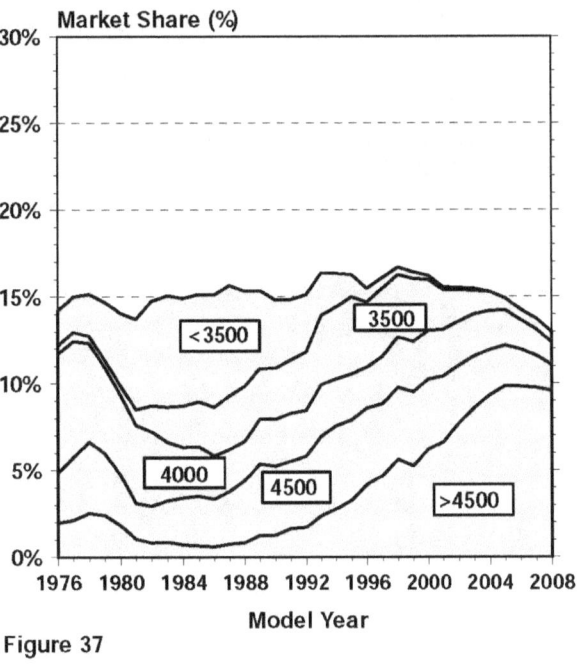

Figure 37

EPA-420-R-09-014   48   November 2009

VI. Fuel Economy Technology Trends

Table 13 repeats the production fraction and adjusted composite fuel economy data from Tables 1 and 2 and adds three measures of powertrain information: engine displacement (CID), horsepower (HP), and specific power (HP/CID). This table also includes production fraction data giving the percent of vehicles that: have front- (FWD) or four-wheel drive (4wd); have manual, lockup, or continuously variable (CVT) transmissions; have port or throttle body fuel injection (TBI) or are Diesels; are equipped with engines that have more than two valves per cylinder; use variable valve timing (VVT); have turbochargers; and use hybrid vehicle technology.

For the overall MY2009 fleet, FWD continues to account for over one-half of the market and 4wd for over one-quarter of the fleet. With transmissions, manuals have dropped to under six percent of the market, while CVTs have grown to eight percent. Nearly 80 percent of the MY2009 fleet has multi-valve engines, and 65 percent use VVT, both all-time highs. Turbochargers are used on about three percent of the fleet. Hybrids represent about two percent of the fleet, while diesels represent 0.5 percent of the projected MY2009 production. Appendix K contains additional data on fuel metering and number of valves per cylinder.

Table 14 compares technology usage for MY2009 by vehicle type and size. As discussed earlier, wheelbase is used in this report to distinguish whether a truck is small, mid-size, or large, and four EPA car classes (Two-Seater, Minicompact, Compact, and Subcompact) have been combined to form the small car class. For this table, the car classes are separated into cars and station wagons, so that the table stratifies light-duty vehicles into a total of 15 vehicle types and sizes. Note that this table does not contain any data for small vans and small pickups, because none have been produced for several years.

Front-wheel drive (FWD) is used heavily in all of the car classes, in small wagons and in midsize vans. Conversely, four-wheel drive (4WD) is used heavily in SUVs and pickups. A large portion of the midsize and large wagons also have 4WD, but very little use of it is made in vans and cars.

Manual transmissions are used primarily in small vehicles and midsize pickups. Similarly, usage of engines with more than two valves per cylinder is more prevalent on small and midsize vehicles than on larger ones.

Detailed tabulations of different technology types, including technology usage percentages for other model years, can be found in the Appendices.

Table 13

Powertrain Characteristics of 1975 to 2009 Light Duty Vehicles (Percentage Basis)

Cars

MODEL YEAR	PROD FRAC	ADJ COMP MPG	ENGINE CID	ENGINE HP	HP/ CID	DRIVETRAIN Front	DRIVETRAIN 4wd	TRANSMISSION Manual	TRANSMISSION Lock	CVT	FUEL METERING GDI	Port	TBI	Dsl	Multi Valve	VVT	TURBO CHRGD	Hybrid
1975	.806	13.5	288	136	.515	6.5		19.6				5.1		.2				
1976	.788	14.9	287	134	.502	5.8		17.1				3.2		.3				
1977	.800	15.6	279	133	.516	6.8		16.8				4.2		.5				
1978	.773	16.9	251	124	.538	9.6		19.8	6.7			5.1		.9				
1979	.778	17.2	238	119	.545	11.9	.3	21.1	8.0			4.7		2.1				
1980	.835	20.0	188	100	.583	29.7	.9	30.9	16.5			6.2	.7	4.4				
1981	.827	21.4	182	99	.594	37.0	.7	29.9	33.3			6.1	2.6	5.9				
1982	.803	22.2	175	99	.609	45.6	.8	29.2	51.4			7.2	9.8	4.7				
1983	.777	22.1	182	104	.615	47.3	3.1	26.1	56.7			9.5	18.9	2.1				
1984	.761	22.4	179	106	.637	53.7	1.0	24.1	58.3			15.0	24.4	1.7				
1985	.746	23.0	177	111	.671	61.6	2.1	22.8	58.7			21.4	32.0	.9				
1986	.717	23.7	167	111	.701	71.1	1.1	24.8	58.0			36.7	28.4	.3	4.8			
1987	.722	23.8	162	112	.732	77.0	1.1	24.9	59.5			42.5	30.5	.3	14.7			
1988	.702	24.1	160	116	.759	81.7	.8	24.3	66.1			53.7	30.0		19.9			
1989	.693	23.7	163	121	.783	82.5	1.0	21.0	69.3	.1		62.4	27.8	.0	24.4			
1990	.698	23.3	163	129	.829	84.6	1.0	19.6	72.9	.0		77.5	21.1	.0	33.0	.6		
1991	.678	23.4	163	132	.851	83.2	1.4	20.5	73.5	.0		78.0	21.8	.1	34.1	2.4		
1992	.666	23.1	170	141	.868	80.8	1.1	17.4	76.4			89.5	10.4	.1	35.0	4.6		
1993	.640	23.5	166	138	.865	85.1	1.2	17.8	77.0			91.6	8.4		36.7	4.8		
1994	.596	23.3	168	143	.884	84.4	.4	16.7	79.3			94.9	5.1		41.0	8.0		
1995	.620	23.4	167	152	.945	82.0	1.2	16.3	81.9			98.8	1.2	.1	52.2	9.8		
1996	.600	23.3	165	154	.958	86.5	1.5	14.9	83.6	.0		98.8	1.1	.1	57.3	11.7	0.3	
1997	.576	23.4	164	156	.974	86.5	1.7	13.5	85.8	.1		99.1	.8	.1	58.6	11.3	0.7	
1998	.551	23.4	164	159	.993	87.0	2.3	12.3	87.3	.1		99.7	.1	.2	61.4	18.4	2.4	
1999	.551	23.0	166	164	1.009	87.2	2.2	10.9	88.4	.0		99.7	.1	.2	64.6	17.1	3.3	
2000	.551	22.9	165	168	1.032	84.9	2.1	11.2	87.7	.0		99.7	.1	.2	65.1	23.4	2.3	.1
2001	.539	23.0	165	168	1.042	84.1	3.2	11.4	87.5	.2		99.7		.3	67.2	28.3	3.6	.0
2002	.515	23.1	166	173	1.066	84.9	3.8	11.2	88.1	.4		99.6		.4	69.9	33.9	4.2	.3
2003	.504	23.2	166	176	1.086	81.7	3.8	11.1	87.9	.9		99.6		.4	73.5	41.2	2.1	.6
2004	.480	23.1	168	182	1.106	80.8	5.4	10.2	88.2	1.4		99.7		.3	77.2	44.2	4.0	.9
2005	.505	23.5	166	182	1.115	79.8	5.8	9.3	88.0	2.6		99.6		.4	78.2	51.6	2.7	2.1
2006	.529	23.3	172	194	1.146	75.8	5.8	9.4	88.1	2.4		99.4		.6	80.8	60.6	3.6	1.5
2007	.529	24.1	165	189	1.157	80.5	5.7	8.5	81.1	10.4		99.7		.0	84.8	66.1	3.7	3.4
2008	.528	24.3	165	193	1.177	77.8	7.3	8.0	80.5	11.5	3.2	96.6		.1	87.9	63.4	4.7	3.4
2009	.513	24.5	167	198	1.195	79.1	6.8	9.1	79.3	11.2	3.9	95.3		.8	90.0	73.9	5.1	2.7

Table 13   (continued)

Powertrain Characteristics of 1975 to 2009 Light Duty Vehicles (Percentage Basis)

Trucks

MODEL YEAR	PROD FRAC	ADJ COMP MPG	ENGINE CID	HP	HP/ CID	DRIVETRAIN Front	4wd	TRANSMISSION Manual	Lock	CVT	FUEL METERING GDI	Port	TBI	Dsl	Multi Valve	VVT	TURBO CHRGD	Hybrid
1975	.194	11.6	311	142	.476		17.1	37.0					.1					
1976	.212	12.2	319	141	.458		22.9	34.8					.1					
1977	.200	13.3	318	147	.482		23.6	32.0					.1					
1978	.227	12.9	314	146	.481		29.0	32.4					.1	.8				
1979	.222	12.5	298	138	.486		18.0	35.2	2.1				.3	1.8				
1980	.165	15.8	248	121	.528	1.4	25.0	53.0	24.6				1.7	3.5				
1981	.173	17.1	247	119	.508	1.9	20.1	51.6	31.1				1.1	5.6				
1982	.197	17.4	243	120	.524	1.7	20.0	45.7	33.2				.7	9.3				
1983	.223	17.8	231	118	.543	1.4	25.8	45.9	36.1				.6	4.7				
1984	.239	17.4	224	118	.557	4.9	31.0	42.1	35.1			1.9	.6	2.3				
1985	.254	17.5	224	124	.586	7.1	30.6	37.1	42.2			8.7	3.5	1.1				
1986	.283	18.2	211	123	.621	5.9	30.3	42.7	42.0			21.8	18.7	.7				
1987	.278	18.3	210	131	.654	7.4	31.5	39.9	44.8			33.3	33.6	.3				
1988	.298	17.9	227	141	.650	9.0	33.3	35.5	53.1			43.3	44.4	.2				
1989	.307	17.6	234	146	.653	9.9	32.0	32.7	56.8			45.9	47.6	.2				
1990	.302	17.4	237	151	.668	15.5	31.3	28.2	67.4			55.2	40.8	.2				
1991	.322	17.8	228	150	.681	9.7	35.3	31.0	67.4			55.0	43.2	.1				
1992	.334	17.4	234	155	.685	13.6	31.4	27.3	71.5			65.9	32.5	.1				
1993	.360	17.5	235	162	.710	15.1	29.4	23.3	75.7			73.4	25.7					
1994	.404	17.2	239	166	.717	13.1	36.9	23.5	75.1			77.2	22.5		5.6			
1995	.380	17.0	244	168	.715	17.7	40.7	20.5	78.6			79.8	20.2		8.4			
1996	.400	17.2	243	179	.757	20.1	37.1	15.6	83.5			99.9		.1	12.4			
1997	.424	17.0	248	187	.775	13.9	43.2	14.6	85.0			100.0		.0	13.7			
1998	.449	17.1	242	187	.795	18.7	42.0	13.4	86.0			100.0		.0	15.8			
1999	.449	16.7	249	197	.814	17.4	44.6	9.1	90.5			100.0			17.3			
2000	.449	16.9	242	197	.832	19.4	42.4	8.0	91.7			100.0			19.9	4.7		
2001	.461	16.7	243	209	.882	18.5	43.8	6.3	93.4			100.0			27.6	9.3		
2002	.485	16.7	244	219	.918	18.5	47.6	5.0	94.7	.0		100.0			35.6	16.2		
2003	.496	16.9	243	221	.927	19.2	46.5	4.8	93.7	1.2		100.0			37.2	19.8	.2	
2004	.520	16.7	252	236	.953	17.2	52.3	3.7	95.0	1.0		100.0			48.4	31.6	.8	
2005	.495	17.2	244	237	.983	25.7	48.3	3.0	95.0	2.0		99.9		.1	52.8	39.8	.6	.1
2006	.471	17.5	240	235	.992	25.1	48.4	3.2	93.5	3.3		99.9		.1	61.4	49.6	.5	1.4
2007	.471	17.7	244	248	1.034	24.9	49.0	2.5	93.9	3.7		99.9		.1	57.0	48.6	1.3	.9
2008	.472	18.2	237	247	1.059	27.8	49.7	2.0	94.1	3.9	1.0	98.6		.2	63.5	52.2	1.1	1.4
2009	.487	18.4	238	253	1.080	29.0	47.6	2.4	92.9	4.7	2.9	96.9		.1	66.3	56.3	1.0	.9

EPA-420-R-09-014

Table 13 (continued)

Powertrain Characteristics of 1975 to 2009 Light Duty Vehicles (Percentage Basis)

Cars and Trucks

MODEL YEAR	PROD FRAC	ADJ COMP MPG	ENGINE CID	ENGINE HP	HP/ CID	DRIVETRAIN Front	DRIVETRAIN 4wd	TRANSMISSION Manual	TRANSMISSION Lock	CVT	FUEL METERING GDI	Port	TBI	Dsl	Multi Valve	VVT	TURBO CHRGD	Hybrid
1975	1.000	13.1	293	137	.507	5.3	3.3	23.2				4.1		.2				
1976	1.000	14.2	294	135	.493	4.6	4.8	20.9				2.5	.0	.2				
1977	1.000	15.1	287	136	.510	5.5	4.7	19.8				3.4	.0	.4				
1978	1.000	15.8	266	129	.525	7.4	6.6	23.0	5.2			3.9	.0	.9				
1979	1.000	15.9	252	124	.532	9.2	4.3	25.1	6.7			3.7	.1	2.0				
1980	1.000	19.2	198	104	.574	25.0	4.9	35.4	17.8			5.2	.8	4.3				
1981	1.000	20.5	193	102	.580	31.0	4.0	34.1	33.0			5.1	2.4	5.9				
1982	1.000	21.1	188	103	.593	37.0	4.6	32.8	47.8			5.8	8.0	5.6				
1983	1.000	21.0	193	107	.599	37.0	8.1	30.8	52.1			7.3	14.8	2.7				
1984	1.000	21.0	190	109	.618	42.1	8.2	28.4	52.8			11.9	18.7	1.8				
1985	1.000	21.3	189	114	.650	47.8	9.3	26.5	54.5			18.2	24.8	.9				
1986	1.000	21.8	180	114	.678	52.6	9.3	29.8	53.5			32.5	25.7	.4				
1987	1.000	22.0	175	118	.710	57.7	9.6	29.1	55.4			39.9	31.4	.3				
1988	1.000	21.9	180	123	.726	60.0	10.5	27.6	62.2			50.6	34.3	.1				
1989	1.000	21.4	185	129	.743	60.2	10.5	24.6	65.5	.1		57.3	33.9	.1				
1990	1.000	21.2	185	135	.781	63.8	10.1	22.2	71.2	.0		70.8	27.0	.1				
1991	1.000	21.2	184	138	.796	59.6	12.3	23.9	71.6	.0		70.6	28.7	.1				
1992	1.000	20.8	191	145	.807	58.4	11.2	20.7	74.8			81.6	17.8	.1				
1993	1.000	20.9	191	147	.809	59.9	11.3	19.8	76.5			85.0	14.6					
1994	1.000	20.4	197	152	.816	55.6	15.2	19.5	77.6			87.7	12.1		26.7			
1995	1.000	20.5	196	158	.857	57.6	16.2	17.9	80.7			91.6	8.4	.0	35.6			
1996	1.000	20.4	197	164	.878	60.0	15.7	15.1	83.5			99.3	.7	.1	39.3		0.2	
1997	1.000	20.1	199	169	.890	55.8	19.3	14.0	85.5	.0		99.5	.5	.1	39.6		0.4	
1998	1.000	20.1	199	171	.904	56.4	20.1	12.8	86.7	.0		99.8	.1	.1	40.9		1.4	
1999	1.000	19.7	203	179	.921	55.8	21.3	10.1	89.4	.0		99.9	.1	.1	43.4		1.8	
2000	1.000	19.8	200	181	.942	55.5	20.2	9.7	89.5	.0		99.8	.0	.1	44.8	15.0	1.3	
2001	1.000	19.6	201	187	.968	53.8	21.9	9.0	90.2	.1		99.9		.1	49.0	19.6	2.0	
2002	1.000	19.4	203	195	.994	52.7	25.0	8.1	91.3	.2		99.8		.2	53.3	25.3	2.2	
2003	1.000	19.6	204	199	1.007	50.7	25.0	8.0	90.8	1.1		99.8		.2	55.5	30.6	1.2	
2004	1.000	19.3	212	211	1.026	47.7	29.8	6.8	91.8	1.2		99.9		.1	62.3	37.6	2.3	.5
2005	1.000	19.9	205	209	1.049	53.0	26.8	6.2	91.4	2.3		99.7		.3	65.6	45.8	1.7	1.1
2006	1.000	20.1	204	213	1.073	51.9	25.8	6.5	90.6	2.8		99.6		.4	71.7	55.4	2.1	1.5
2007	1.000	20.6	203	217	1.099	54.3	26.1	5.2	87.1	7.2		99.8		.1	71.7	57.9	2.6	2.2
2008	1.000	21.0	199	219	1.122	54.2	27.3	5.2	86.9	7.9	2.2	97.5		.1	76.4	58.1	3.0	2.5
2009	1.000	21.1	202	225	1.139	54.7	26.7	5.7	86.0	8.1	3.5	96.1		.5	78.5	65.3	3.1	1.8

EPA-420-R-09-014

Table 14

MY2009 Technology Usage by Vehicle Type and Size
(Percent of Vehicle Type/Size Strata)

Vehicle Type	Size	Front Wheel Drive	Four Wheel Drive	Manual Trans	Multi-Valve	Variable Valve
Car	Small	77%	6%	16%	92%	69%
	Midsize	86%	6%	4%	97%	84%
	Large	75%	3%	0%	72%	69%
	All	79%	5%	8%	89%	74%
Wagon	Small	87%	13%	18%	100%	82%
	Midsize	40%	60%	5%	100%	33%
	Large	0%	100%	0%	100%	88%
	All	77%	22%	16%	100%	73%
Van	Small					
	Midsize	98%	2%	0%	61%	40%
	Large	0%	14%	0%	0%	
	All	94%	2%	0%	59%	39%
SUV	Small	0%	94%	24%	35%	4%
	Midsize	35%	55%	2%	86%	76%
	Large	29%	52%	0%	72%	57%
	All	31%	55%	2%	77%	64%
Pickup	Small					
	Midsize	0%	34%	19%	90%	45%
	Large	0%	48%	1%	30%	45%
	All	0%	46%	4%	40%	42%

EPA-420-R-09-014   November 2009

Figures 38 through 42 show trends in drive use for the five vehicle classes based on 3-year moving averages. Cars used to be nearly all rear-wheel drive, but have been 80+ percent front-wheel drive since the late 1980s. Only a small percentage of wagons still have rear-wheel drive, but in recent years they have made substantial use of 4WD.

The trend towards increased use of front wheel drive for vans is very similar to that for cars, except it started a few years later and appears to be continuing. Over 90 percent of vans currently use front-wheel drive, compared to essentially none before 1984, which coincides with the introduction of minivans to the U.S. market. SUVs are mostly 4WD; but a trend toward front-wheel drive SUVs started in MY2000. Pickups remain the bastion of rear-wheel drive with the increasing amount of 4WD the only other drive option. Except for a brief period in the early 1980s, front-wheel drive has not been used in pickups.

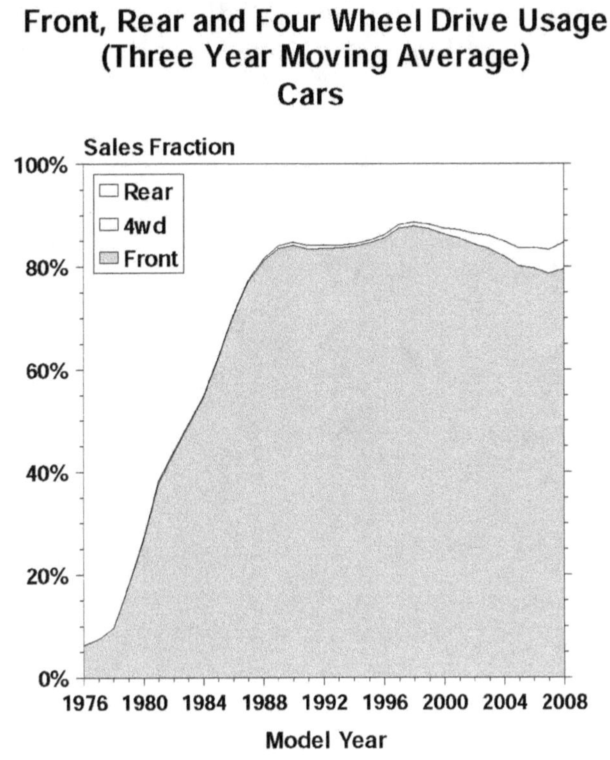

Front, Rear and Four Wheel Drive Usage (Three Year Moving Average) Cars

Figure 38

Front, Rear and Four Wheel Drive Usage (Three Year Moving Average) Wagons

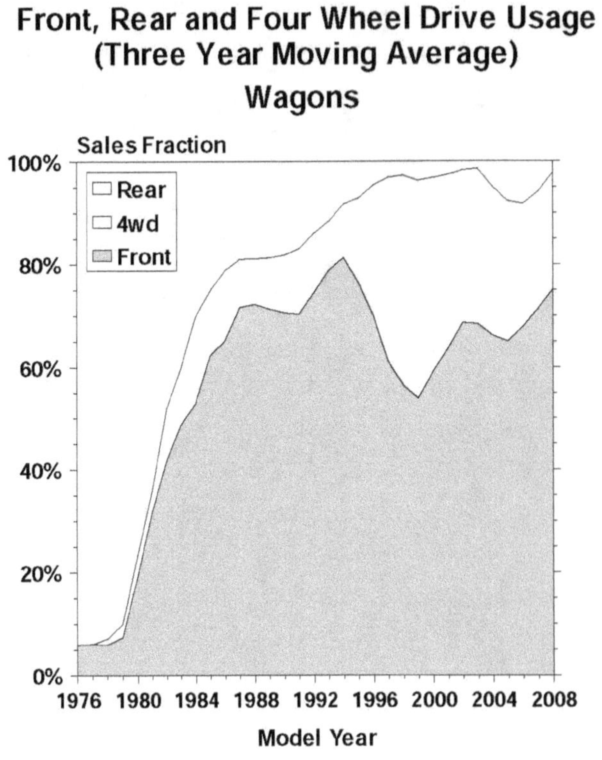

Figure 39

EPA-420-R-09-014   54   November 2009

Front, Rear and Four Wheel Drive Usage (Three Year Moving Average) Vans

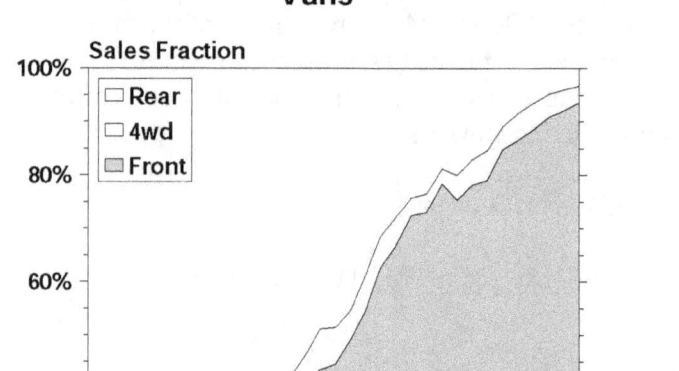

Figure 40

Front, Rear and Four Wheel Drive Usage (Three Year Moving Average) SUVs

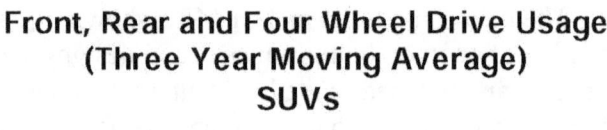
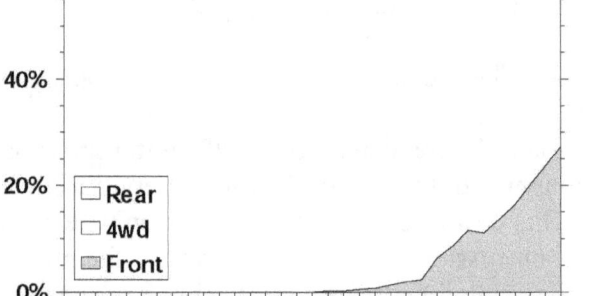

Figure 41

Front, Rear and Four Wheel Drive Usage (Three Year Moving Average) Pickups

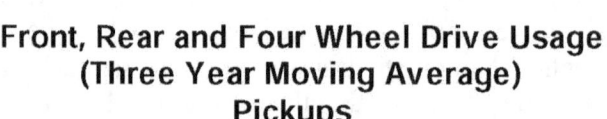
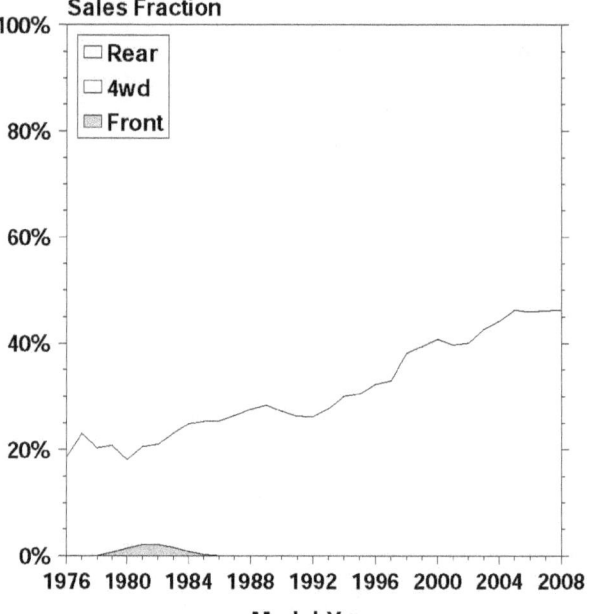

Figure 42

EPA-420-R-09-014   55   November 2009

The increasing trend in Ton-MPG shown in Table 1 can be attributed to better vehicle design, including more efficient engines, better transmission designs, and better matching of the engine and transmission. Powertrains are matched to the load better when the engine operates closer to its best efficiency point more often. For many conventional engines, this point is approximately 2000 RPM and two-thirds of the maximum torque at that speed. One way to make the engine operate more closely to its best efficiency point is to increase the number of gears in the transmission and, for automatic transmissions, employing a lockup torque converter. Three important changes in transmission design have occurred in recent years:

1. The use of additional gears for both automatic and manual transmissions;

2. For the automatics, conversion to lockup (L3, L4, L5, L6, and now L7) torque converter transmissions; and

3. The use of continuously variable transmissions (CVTs).

Table 15 compares Ton-MPG by transmission and vehicle type for 1988, the peak year for passenger car fuel economy, and this year. In 1988, every transmission type shown in the table achieved less than 40 Ton-MPG. This year, nearly every transmission type achieves at least 40 Ton-MPG. Figures 43 to 46 indicate that the L4 transmission is losing its position as the predominant transmission type for all vehicle classes. Use of the L4 transmission for cars peaked at about 80 percent in 1999 and is now down to about 40 percent. Similarly, its use peaked at over 90 percent in 1996 for SUVs and has dropped to about 25 percent. Over half of this year's pickups will still have L4 transmissions. Where manual transmissions are used, the 5-speed (M5) transmission now predominates.

Transmissions alter the ratio of engine speed to drive wheel speed. In conventional transmissions, this speed ratio is limited to a fixed number of discrete values, but for a CVT, the ratio is continuous. These transmissions differ from conventional automatic transmissions and manual transmissions in that CVTs do not have a fixed number of gears with the advantage that the engine speed/drive wheel speed ratio can be altered to enhance vehicle performance or fuel economy in ways not available with conventional transmissions.

More data stratified by transmission type can be found in Appendix I.

EPA-420-R-09-014   56   November 2009

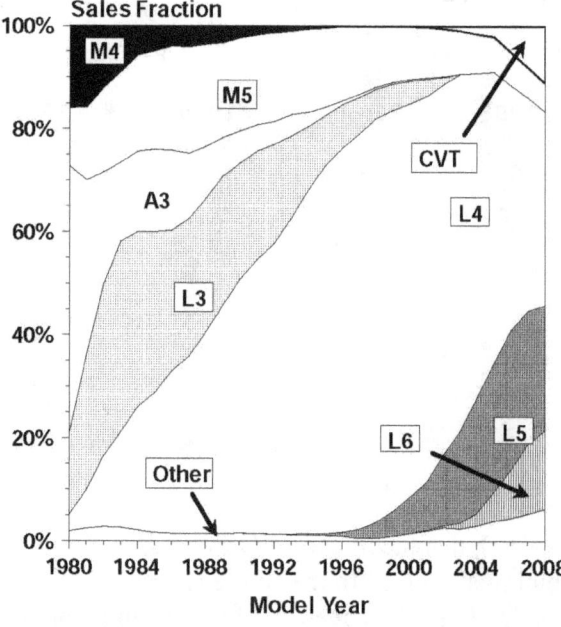

**Transmission Sales Fraction
(Three Year Moving Average)
Cars**

Figure 43

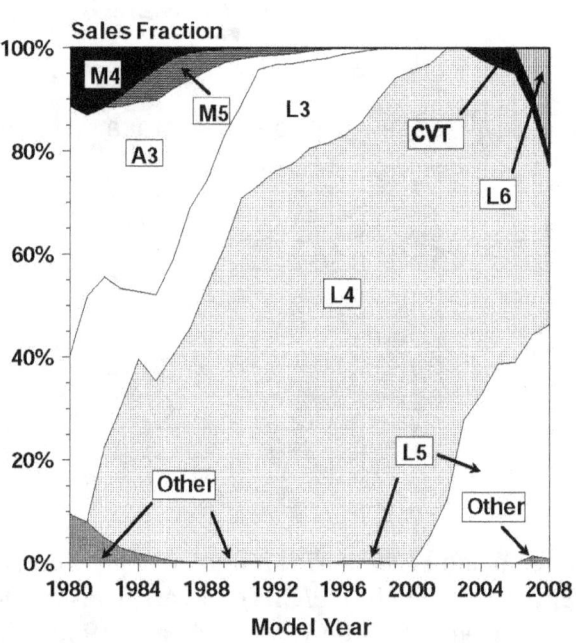

**Transmission Sales Fraction
(Three Year Moving Average)
Vans**

Figure 44

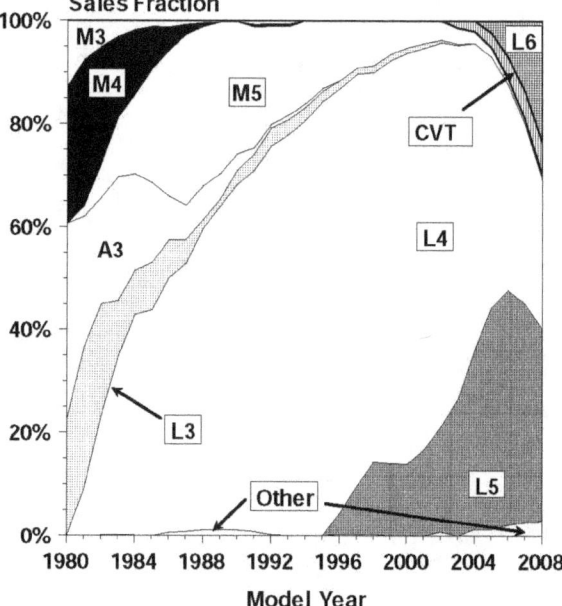

**Transmission Sales Fraction
(Three Year Moving Average)
SUVs**

Figure 45

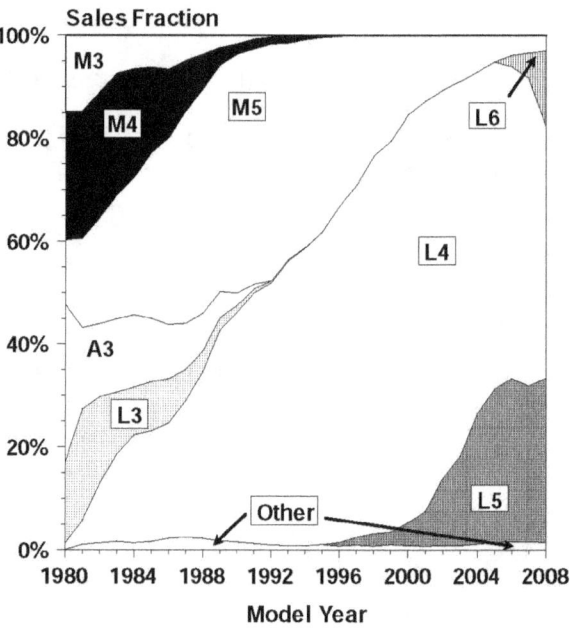

**Transmission Sales Fraction
(Three Year Moving Average)
Pickups**

Figure 46

EPA-420-R-09-014   57   November 2009

Table 15

Ton-MPG by Transmission and Vehicle Type

(Conventionally Powered Vehicles)

Trans	Car 1988	Car 2009	Van 1988	Van 2009	SUV 1988	SUV 2009	Pickup 1988	Pickup 2009
M4	37.0	--	33.6	--	38.0	--	32.4	--
M5	37.7	41.7	37.7	--	33.1	42.3	35.3	40.3
M6	--	39.6	--	--	--	36.8	--	38.1
CVT	--	44.6	--	--	--	43.0	--	--
L3	36.1	--	37.1	--	33.5	--	31.4	--
L4	37.9	42.4	36.6	44.7	33.8	41.3	33.8	42.7
L5	--	44.5	--	45.9	--	42.1	--	41.3
L6	--	43.3	--	45.7	--	45.2	--	46.0

Table 16 and Figures 47 through 50 compare horsepower (HP), displacement (CID), and specific power or horsepower per cubic inch (HP/CID) for cars, vans, SUVs, and pickups. For all four vehicle types, significant CID reductions occurred in the late 1970s and early 1980s. Engine displacement has been flat for cars and vans since the mid-1980s and has declined slightly for SUVs since the mid-1990s, but has been increasing for two decades for pickups. Average horsepower has increased substantially for all of these vehicle types since 1981 with the highest increase occurring for pickups whose HP is now more than double what it was then (i.e., 282 versus 115 HP). Light-duty vehicle engines, thus, have also improved in specific power with the highest specific power being for engines used in passenger cars.

Table 16

MY2009 Engine Characteristics by Vehicle Type

Vehicle Type	HP	CID	HP/ CID	Multi- Valve	Variable Valve	Cylinder Deactivation
Car	198	167	1.20	90%	74%	3%
Van	221	223	1.00	59%	39%	18%
SUV	247	221	1.13	78%	64%	11%
Pickup	282	289	0.98	40%	42%	26%
All	225	202	1.14	79%	65%	9%

EPA-420-R-09-014

Car Horsepower, CID and Horsepower per CID (Three Year Moving Average)

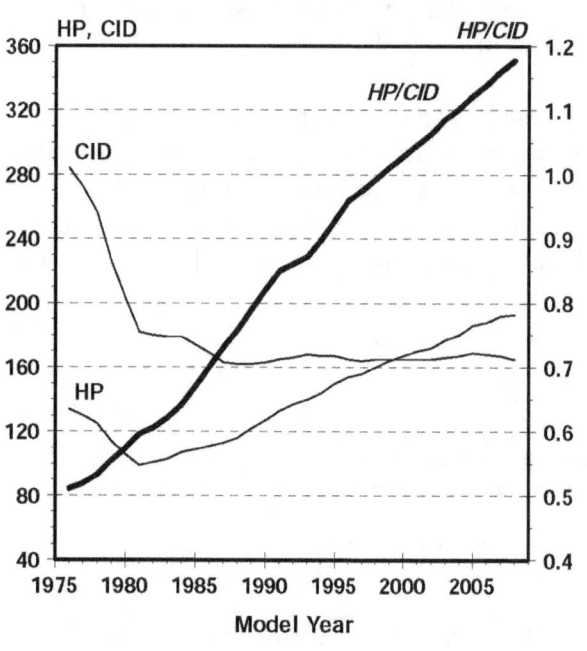

Figure 47

Van Horsepower, CID and Horsepower per CID (Three Year Moving Average)

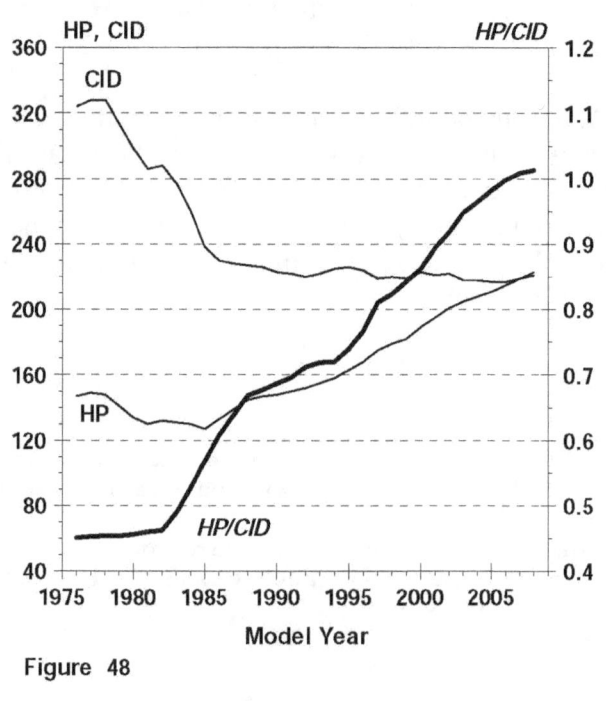

Figure 48

SUV Horsepower, CID and Horsepower per CID (Three Year Moving Average)

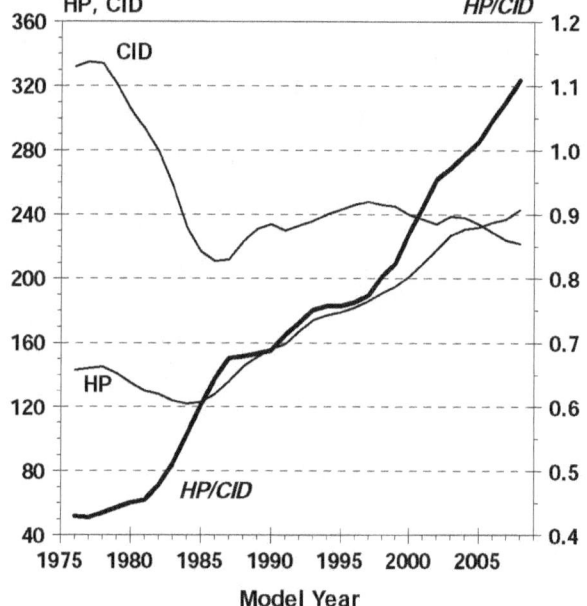

Figure 49

Pickup Horsepower, CID and Horsepower per CID (Three Year Moving Average)

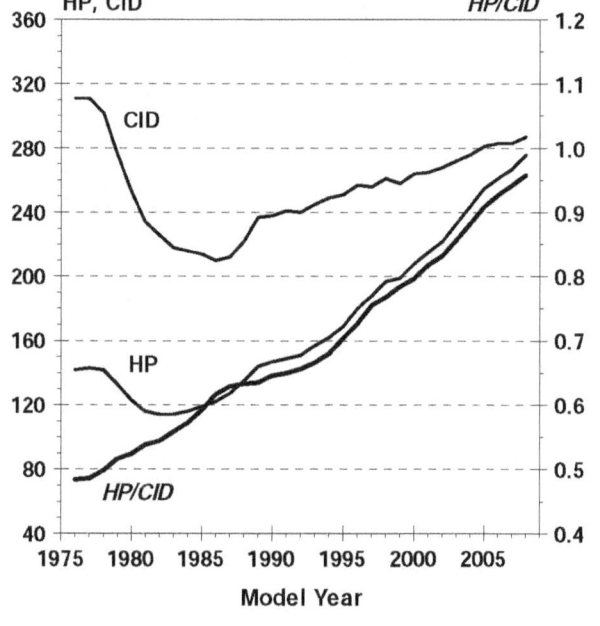

Figure 50

EPA-420-R-09-014   November 2009

Table 17 compares CID, HP, and HP/CID by vehicle type and number of cylinders for model years 1988 and 2009. Table 17 shows that the increase in horsepower shown for the fleet in Table 13 extends to all vehicle type and cylinder number strata. These increases in horsepower range from 47 to 99 percent. Because displacement has remained relatively constant, it can be seen that the primary reason for the horsepower increase is increased specific power - up between 38 and 102 percent from 1988 to 2009.

At the number-of-cylinders level of stratification, model year 2009 cars generally achieve higher specific power than vans, SUVs, or pickups. One reason for the lower specific power of some truck engines is that these vehicles may be used to carry heavy loads or pull trailers and thus need more "torque rise," (i.e., an increase in torque as engine speed falls from the peak power point) to achieve acceptable drivability. Engines equipped with four valves per cylinder typically have inherently lower torque rise than two valve engines with lower specific power.

Table 17

Changes in Horsepower and Specific Power
by Vehicle Type and Number of Cylinders

Vehicle Type	Cyl.	HP 1988	HP 2009	Percent Change	CID 1988	CID 2009	Percent Change	HP/CID 1988	HP/CID 2009	Percent Change
Cars	4	95	153	61%	118	127	8%	0.805	1.201	49%
	6	142	251	77%	193	209	8%	0.744	1.209	63%
	8	164	327	99%	301	299	-1%	0.544	1.098	102%
Vans	6	149	219	47%	213	221	4%	0.722	0.996	38%
	8	168	301	79%	322	325	1%	0.520	0.926	78%
SUVs	4	94	173	84%	122	144	18%	0.773	1.208	56%
	6	147	248	69%	211	218	3%	0.706	1.142	62%
	8	183	324	77%	338	322	-5%	0.541	1.008	86%
Pickups	4	97	157	62%	142	157	11%	0.685	0.999	46%
	6	142	230	62%	229	239	4%	0.644	0.967	50%
	8	180	314	74%	329	322	-2%	0.544	0.971	79%

Table 18

Changes in Horsepower and Specific Power
by Vehicle Type and Inertia Weight

Cars

Inertia Weight	HP 1988	HP 2009	Percent Change	CID 1988	CID 2009	Percent Change	HP/CID 1988	HP/CID 2009	Percent Change
2000	59	70	19%	77	61	-21%	0.770	1.148	49%
2250	73	225	208%	90	110	22%	0.808	2.045	153%
2500	78	106	36%	100	91	-9%	0.785	1.165	48%
2750	97	123	27%	123	105	-15%	0.804	1.179	47%
3000	114	138	21%	145	117	-19%	0.797	1.174	47%
3500	151	182	21%	212	151	-29%	0.732	1.216	66%
4000	160	255	59%	289	216	-25%	0.569	1.201	111%
4500	144	316	119%	305	282	-8%	0.474	1.135	140%
5000	207	406	96%	408	318	-22%	0.509	1.287	153%
5500	205	320	56%	412	250	-39%	0.498	1.271	155%
6000	205	523	155%	412	350	-15%	0.498	1.472	196%

Vans

Inertia Weight	HP 1988	HP 2009	Percent Change	CID 1988	CID 2009	Percent Change	HP/CID 1988	HP/CID 2009	Percent Change
4500	169	216	28%	320	218	-32%	0.528	0.996	89%
5000	156	244	56%	312	244	-22%	0.500	1.001	100%
5500	195	301	54%	346	325	-6%	0.562	0.926	65%
6000	126	301	139%	379	325	-14%	0.332	0.926	179%

SUVs

Inertia Weight	HP 1988	HP 2009	Percent Change	CID 1988	CID 2009	Percent Change	HP/CID 1988	HP/CID 2009	Percent Change
3500	147	172	17%	210	146	-30%	0.712	1.175	65%
4000	135	204	51%	190	179	-6%	0.723	1.156	60%
4500	147	250	70%	311	221	-29%	0.494	1.133	129%
5000	181	274	51%	330	244	-26%	0.545	1.139	109%
5500	200	322	61%	350	298	-15%	0.572	1.101	92%
6000	162	331	104%	368	326	-11%	0.445	1.019	129%

Pickups

Inertia Weight	HP 1988	HP 2009	Percent Change	CID 1988	CID 2009	Percent Change	HP/CID 1988	HP/CID 2009	Percent Change
3500	129	156	21%	183	154	-16%	0.719	1.010	40%
4000	154	210	36%	282	217	-23%	0.555	0.971	75%
4500	174	243	40%	322	240	-25%	0.539	1.019	89%
5000	193	245	27%	342	274	-20%	0.565	0.897	59%
5500	178	317	78%	363	323	-11%	0.495	0.980	98%
6000	140	335	139%	379	333	-12%	0.369	1.003	171%

Table 18 shows similar data to that in Table 17, but the stratification is based on inertia weight. This table clearly shows that, for every case for which a comparison can be made between 1988 and 2009, there were increases in HP, substantial increases in specific power ranging from 40 to 196 percent, and with just minor exceptions, substantial decreases in CID.

EPA-420-R-09-014   November 2009

HP/CID by Number of Valves Per Cylinder
(Three Year Moving Average)
Cars

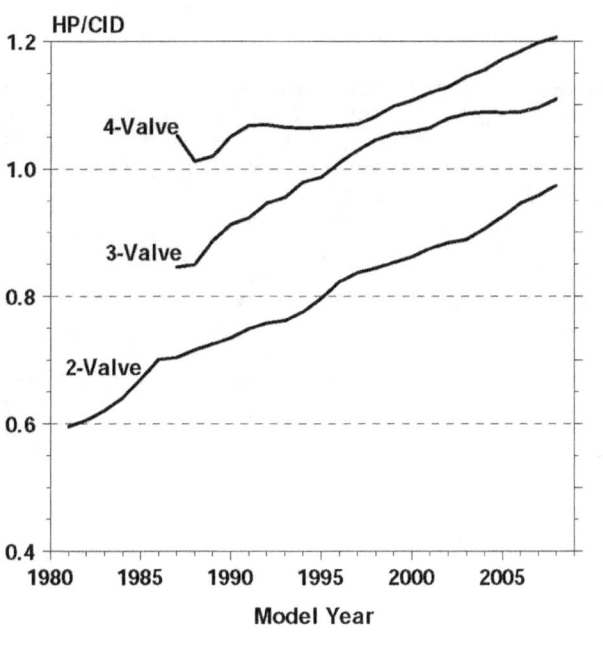

Figure 51

HP/CID by Number of Valves Per Cylinder
(Three Year Moving Average)
Vans

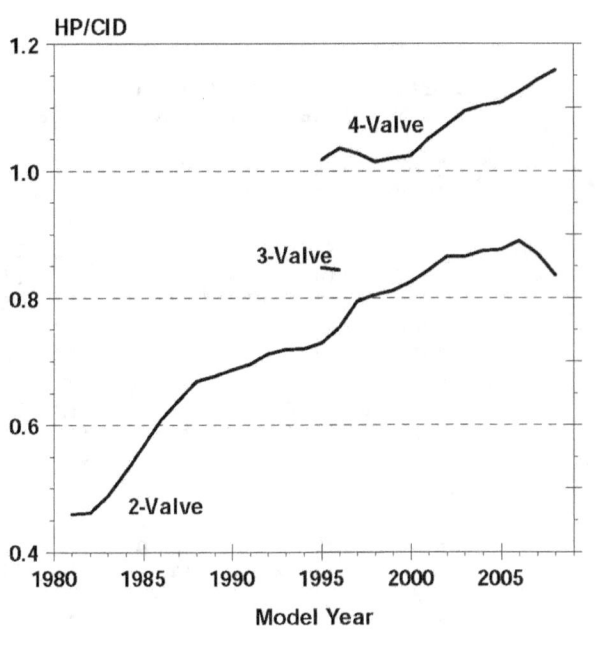

Figure 52

HP/CID by Number of Valves Per Cylinder
(Three Year Moving Average)
SUVs

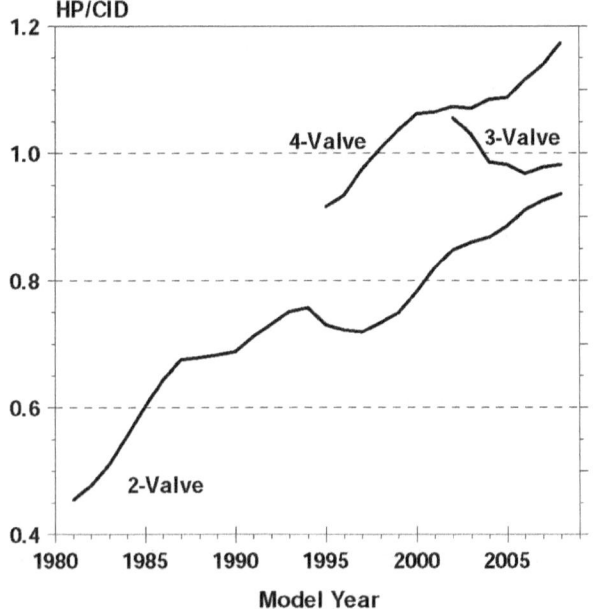

Figure 53

HP/CID by Number of Valves Per Cylinder
(Three Year Moving Average)
Pickups

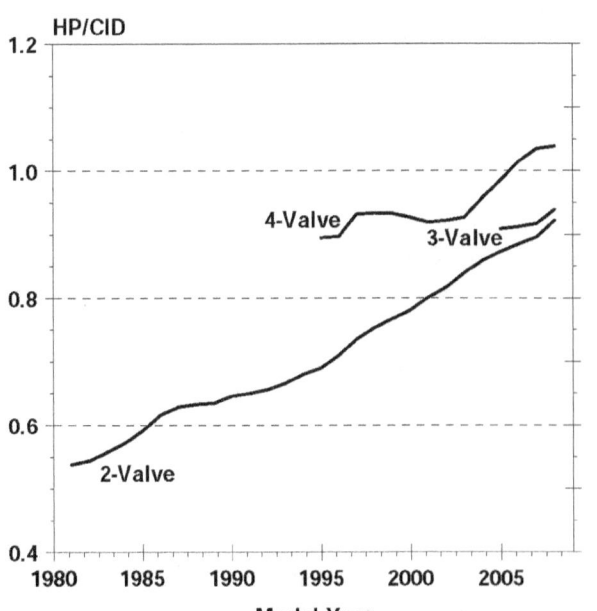

Figure 54

EPA-420-R-09-014   62   November 2009

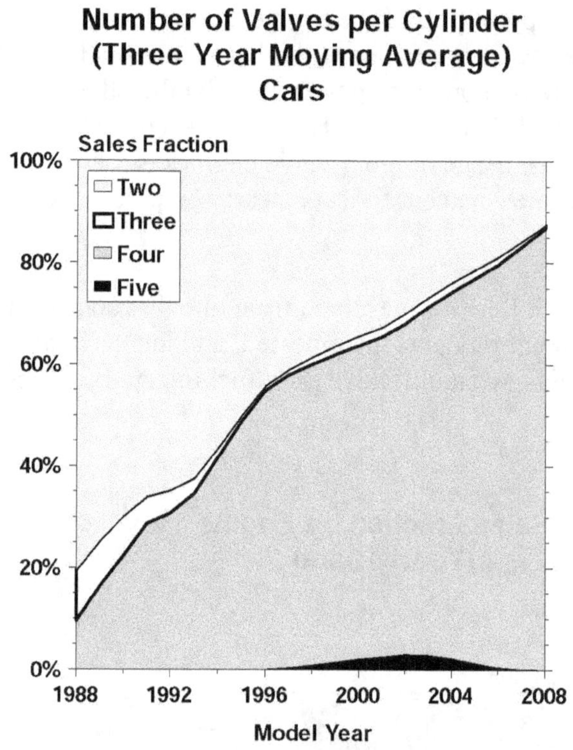

Number of Valves per Cylinder (Three Year Moving Average) Cars

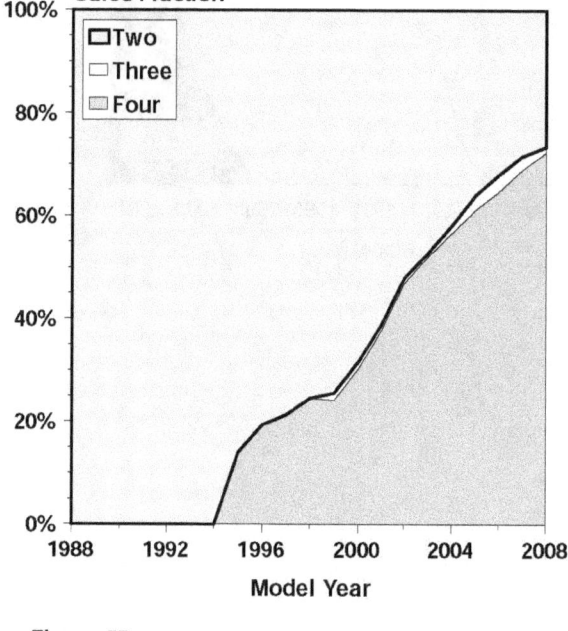

Figure 55

Number of Valves per Cylinder (Three Year Moving Average) Vans

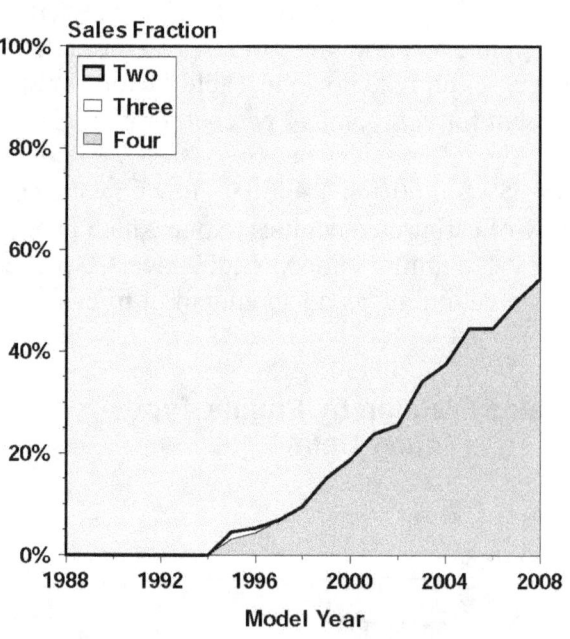

Figure 56

Number of Valves per Cylinder (Three Year Moving Average) SUVs

Figure 57

Number of Valves per Cylinder (Three Year Moving Average) Pickups

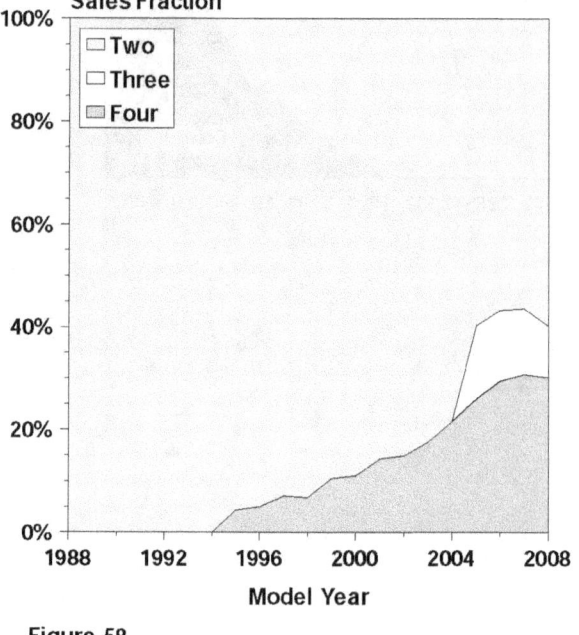

Figure 58

EPA-420-R-09-014   63   November 2009

Figures 51 through 54 show that increases in HP per CID apply to all of the engines, except for a couple of cases for engines with three valves. Engines with more valves per cylinder deliver higher values of HP per CID. Engines with *only* two valves per cylinder deliver substantially more horsepower per CID then they used to, typically a 50 – 80 percent increase for the time period shown. The increases in HP and HP-per-CID are due to changes in engine technologies. Figures 55 through 58 show that usage of multi-valve engines is increasing for all vehicle types and as shown in Table 16 for MY2009, is now 90 percent for cars, nearly 80 percent for SUVs, 60 percent for vans, and 40 percent for pickups.

Figures 59 and 60 and Table 19 show how the car and truck fleet have evolved from one that consisted almost entirely of carbureted engines to one which is now almost entirely port fuel injected, and increasingly using variable valve timing. For MY2009, over 70 percent of cars have multi-valve, port fuel injected engines with variable valve timing, as do about half of trucks.

Car Sales Fraction by Engine Type
(Yearly Data)

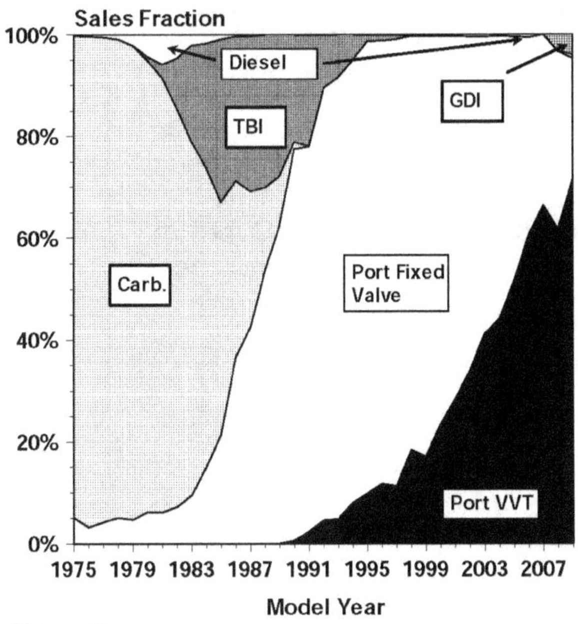

Figure 59

Truck Sales Fraction by Engine Type
(Yearly Data)

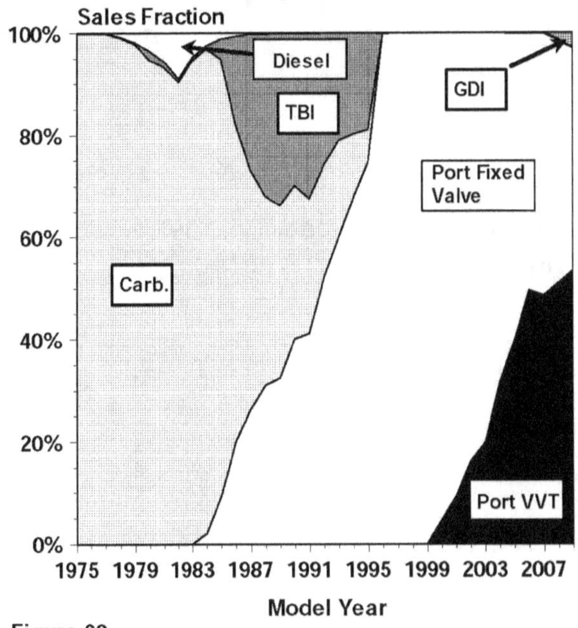

Figure 60

Table 19

Production Fraction of MY1988 and MY2009 Light Vehicles
by Engine Type and Valve Timing

Engine Type	Cars		Vans		SUVs		Pickups		All	
	1988	2009	1988	2009	1988	2009	1988	2009	1988	2009
Carb	16%	---	<1%	---	16%	---	16%	---	15%	---
TBI	30%	---	43%	---	37%	---	48%	---	34%	---
Port Fixed	54%	24%	57%	61%	47%	36%	36%	58%	51%	33%
Port Variable	---	69%	---	39%	---	59%	---	42%	---	61%
GDI Fixed	---	1%	---	---	---	---	---	---	---	1%
GDI Variable	---	2%	---	---	---	4%	---	---	---	3%
Diesel	<1%	1%	<1%	---	<1%	<1%	<1%	---	<1%	<1%
Hybrids	---	3%	---	---	---	1%	---	<1%	---	2%

For many years, automotive manufacturers have been using engines which use either cams or electric solenoids to provide variable intake and/or exhaust valve timing and in some cases valve lift. Conventional engines use camshafts which are permanently synchronized with the engine's crankshaft so that they operate the valves at a specific fixed point in each combustion cycle regardless of the speed and load at which the engine is operated. The ability to control valve timing allows the design of an engine combustion chamber with a higher compression level than in engines equipped with fixed valve timing engines which in turn provides greater engine efficiency, more power and improved combustion efficiency. Variable valve timing (VVT) also allows the valves to be operated at different points in the combustion cycle, to provide performance that is precisely tailored to the engine's specific speed and load at any given instant with the valve timing set to allow the best overall performance across the engine's normal operating range. This results in improved engine efficiency under low-load conditions, such as at idle or highway cruising, and increased power at times of high demand. In addition, variable valve timing can result in reduced pumping losses, from the work required to pull air in and push exhaust out of the cylinder.

Because automobile manufacturers are not currently required to provide EPA with data on the type of valve timing their engines have, the data base used to generate EPA's fuel economy trend report was augmented to indicate whether a vehicle had fixed or variable valve timing. The data augmentation was based on data from trade publications and data published by automotive manufacturers. In addition, no differentiation between engines which used cams or solenoids to control the valve timing was made, nor was valve lift considered. For cars, the augmented data covers model years 1989 to 2009, while for trucks the augmentation covered model years 1999 to 2009.

Table 20

Comparison of MY1988 and MY2009 Cars
by Engine Fuel Metering, Number of Valves and Valve Timing

Fuel Metering	Number of Valves	Valve Timing	Horsepower		CID		HP/CID		Ton MPG		0 to 60 Time	
			1988	2009	1988	2009	1988	2009	1988	2009	1988	2009
Carb		Fixed	88	---	131	---	.75	---	37.2	---	14.3	---
TBI	2	Fixed	97	---	141	---	.71	---	36.9	---	13.7	---
Port	2	Fixed	136	269	193	286	.74	.96	36.6	40.5	11.9	8.9
Port	4	Fixed	137	198	131	172	1.05	1.16	37.9	41.4	11.1	9.6
Port	4	Variable	---	190	---	154	---	1.23	---	43.5	---	9.5
GDI	4	Fixed	---	204	---	121	---	1.68	---	44.3	---	9.6
GDI	4	Variable	---	275	---	184	---	1.54	---	43.7	---	7.8

Percent Change over 1988 Port Two Valve, Fixed Valve Timing

Fuel Metering	Number of Valves	Valve Timing	Horsepower		CID		HP/CID		Ton MPG		0 to 60 Time	
Carb		Fixed	-35%	---	-32%	---	1%	---	2%	---	20%	---
TBI	2	Fixed	-29%	---	-27%	---	-4%	---	1%	---	15%	---
Port	2	Fixed	0%	98%	0%	48%	0%	30%	0%	11%	0%	-25%
Port	4	Fixed	1%	46%	-32%	-11%	42%	57%	4%	13%	-7%	-19%
Port	4	Variable	--	40%	---	-20%	---	66%	---	19%	---	-20%
GDI	4	Fixed	---	50%	---	-37%	---	127%	---	21%	---	-19%
GDI	4	Variable	---	102%	---	-5%	---	108%	---	19%	---	-34%

Table 20 compares horsepower, engine size (CID), specific power (HP/CID), Ton- mpg, and estimated 0-to-60 acceleration time for five selected MY1988 and 2009 engine types.

Because 1988 was the peak year for car fuel economy, and because the two valve, fixed valve timing, port injected engine accounted for about half of the car engines built that year, it was selected as a baseline engine with its average characteristics compared to those for the MY2009 two- and four-valve, fixed valve timing and four-valve VVT engines. As shown in Figure 61, all three of these MY2009 engine types had substantially higher horsepower than the baseline MY1988 engine, but the MY2009 four valve engines fixed and VVT engines are considerably smaller and have substantially higher specific power. Not all of these improvements in engine design for these engine types that occurred between 1988 and 2009 were used to improve fuel economy as indicated by the nominal 20 percent decrease in 0-to-60 time each achieved. As mentioned earlier, in this report vehicle performance for conventionally powered vehicles is determined by an estimate of 0-to-60 acceleration time calculated from the ratio of vehicle power to weight. Obtaining increased power to weight in a time when weight is trending upwards implies that horsepower is increasing. Increased horsepower can be obtained by increasing the engine's displacement, the engine's specific power (HP/CID), or both. Increasing specific power has been the primary driver for increases in performance for the past two decades.

EPA-420-R-09-014   66   November 2009

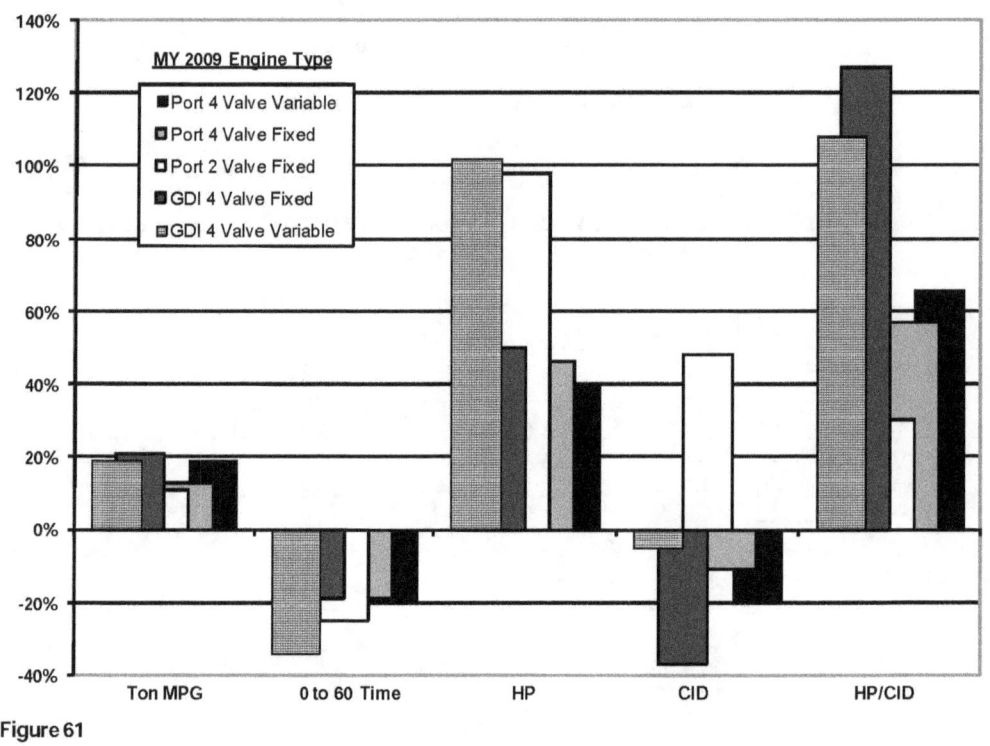

Figure 61

For the current model year fleet, specific power has been studied at an even more detailed level of stratification with both car and truck engines being classified according to: (1) the number of valves per cylinder, (2) the manufacturer's fuel recommendation, (3) the presence or absence of an intake boost device such as a turbocharger or supercharger, and (4) whether or not the engine had fixed or variable valve timing (see Tables 21 and 22). Higher HP/CID is associated with: (a) more valves per cylinder, (b) higher octane fuel, (c) intake boost, and (d) use of variable valve timing. The technical approaches result in specific power ranges for cars and trucks from about .9 to about 1.8. The relative production fractions in Tables 21 and 22 are just for each technical option in the table and exclude hybrids.

Tables 21 and 22 show the incremental effect, on a production weighted basis, of adding each technical option, but not all of the technical options are production significant. The effect of the use of higher octane fuel cannot be discounted, because roughly 18 percent of the current car fleet is comprised of vehicles which use engines for which high octane fuel is recommended. By comparison, about 11 percent of this year's light trucks require premium fuel.

Engine technology which delivers improved specific power thus can be used in many ways ranging from reduced displacement and improved fuel economy at constant (or worse) performance, to increased performance and the same fuel economy at constant displacement.

EPA-420-R-09-014   67   November 2009

Table 21

HP/CID and Production Fraction by Fuel and Engine Technology

Model Year 2009 Cars

Number of Valves per Cylinder

Fuel/Boost/Valves	Two		Three		Four		Five		Total
	HP/CID	Prod Fract.	HP/CID	Prod Fract.	HP/CID	Prod Fract.	HP/CID	Prod Fract.	Prod Fract.
Regular/No Boost/FIX	.88	.029	----	----	1.14	.183	----	----	.212
Regular/No Boost/VVT	1.02	.066	1.07	.004	1.17	.522	1.35	----	.592
Regular/Boost /FIX	----	----	----	----	1.72	.002	----	----	.002
Regular/Boost /VVT	----	----	----	----	1.76	.007	----	----	.007
Premium/No Boost/FIX	1.47	.005	1.32	----	1.22	.019	1.35	----	.024
Premium/No Boost/VVT	----	----	1.37	----	1.36	.118	----	----	.119
Premium/Boost /FIX	1.65	----	1.69	----	1.68	.015	----	----	.015
Premium/Boost /VVT	----	----	1.18	.001	1.70	.021	----	----	.021
Diesel/No Boost	----	----	----	----	----	----	----	----	.000
Diesel/Boost	----	----	----	----	1.17	.008	----	----	.008
Other	----	----	----	----	----	----	----	----	.000
Total		.100		.005		.895		----	1.000

Table 22

HP/CID and Production Fraction by Fuel and Engine Technology

Model Year 2009 Trucks

Number of Valves per Cylinder

Fuel/Boost/Valves	Two		Three		Four		Five		Total
	HP/CID	Prod Fract.	HP/CID	Prod Fract.	HP/CID	Prod Fract.	HP/CID	Prod Fract.	Prod Fract.
Regular/No Boost/FIX	.91	.282	1.04	.008	1.12	.137	----	----	.426
Regular/No Boost/VVT	1.01	.053	.94	.013	1.15	.399	----	----	.464
Regular/Boost /FIX	----	----	----	----	----	----	----	----	.000
Regular/Boost /VVT	----	----	----	----	1.56	----	----	----	.000
Premium/No Boost/FIX	----	----	1.15	.001	1.17	.008	----	----	.008
Premium/No Boost/VVT	1.14	.001	----	----	1.24	.088	1.38	----	.089
Premium/Boost /FIX	----	----	1.53	----	----	----	----	----	.000
Premium/Boost /VVT	----	----	----	----	1.64	.010	----	----	.010
Diesel/No Boost	----	----	----	----	----	----	----	----	.000
Diesel/Boost	----	----	----	----	1.17	.001	----	----	.001
Other	----	----	----	----	----	----	----	----	.000
Total		.336		.021		.642		----	1.000

EPA-420-R-09-014

A relatively recent engine development has been the reintroduction of cylinder deactivation, an automotive technology that was used by General Motors in some MY1981 V-8 engines that could be operated in 8- , 6- and 4-cylinder modes. This approach, which has also been called by a number of names including 'variable displacement', 'displacement on demand', 'active fuel management' and 'multiple displacement', involves allowing the valves of selected cylinders of the engine to remain closed and interrupting the fuel supply to these cylinders when engine power demands are below a predetermined threshold, as typically happens under less demanding driving conditions, such as steady state operation. Under light load conditions, the engine can thus provide better fuel mileage than would otherwise be achieved. Although frictional and thermodynamic energy losses still occur in the cylinders that are not being used, these losses are more than offset by the increased load and reduced specific fuel consumption of the remaining cylinders. Typically half of the usual number of cylinders are deactivated. Challenges to the engine designer for this type of engine include mode transitions, idle quality, and noise and vibration. For MY2009, as shown previously in Table 16, it is estimated that about nine percent of all vehicles are equipped with cylinder deactivation.

Table 23 compares three examples of individual MY2009 car models with and without cylinder deactivation. Table 24 shows two truck cases as well. The Honda Odyssey is the only model shown that offers the same engine with and without cylinder deactivation. In this case, cylinder deactivation increases fuel economy by eight percent. For the two cases shown where cylinder deactivation is offered with a smaller, less powerful engine, this combination led to about 25 percent higher fuel economy relative to the larger engine without cylinder deactivation. In the two cases shown where cylinder deactivation was coupled with a larger, more powerful engine, this combination led to 4-9 percent lower fuel economy compared to the smaller engine.

Table 23

Comparison of MY2009 Cars with Engines with Cylinder Deactivation

Car Class	Model Name	Drive	Trans	Inertia Weight	Engine CID	HP	Lab. 55/45	Cyl. Deact.	Pct. HP	Change MPG
Small Car	Challenger	Rear	L5	4500	348	340	24.1	Yes	-20%	23%
	Challenger				372	425	19.6	No		
Midsize Car	Lacrosse-Allure	Front	L4	4000	325	290	24.7	Yes	45%	-9%
	Lacrosse-Allure				231	200	27.1	No		
Large Car	300 AWD	4wd	L5	4500	348	340	23.6	Yes	36%	-4%
	300 AWD				215	250	24.6	No		

Table 24

Comparison of MY2009 Trucks with Engines with Cylinder Deactivation

Truck Class	Model Name	Drive	Trans	Inertia Weight	Engine CID	HP	Lab. 55/45	Cyl. Deact.	Pct. HP	Change MPG
Midsize Van	Odyssey	Front	L5	4500	212	241	25.9	Yes	-1%	8%
	Odyssey				212	244	23.9	No		
Large SUV	Trailblazer	4WD	L4	5000	325	300	21.1	Yes	-23%	25%
	Trailblazer				364	390	16.9	No		

Car Technology Penetration
Years After First Significant Use

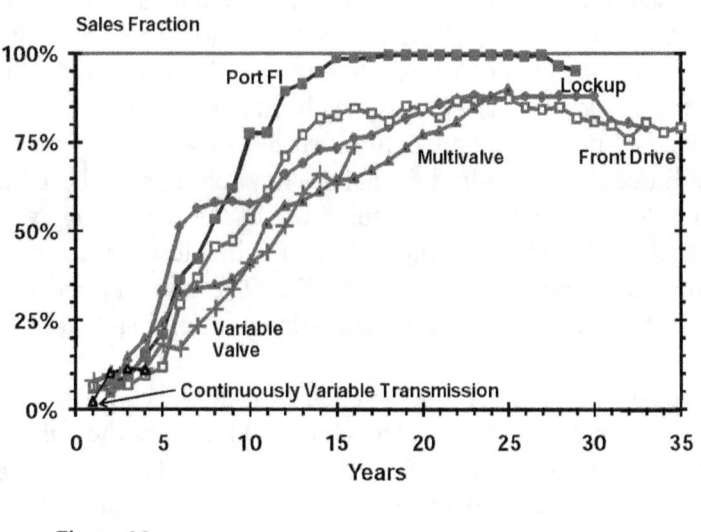

Figure 62

Figure 62 compares penetration rates for six passenger car technologies, namely port fuel injection (Port FI), front-wheel drive (FWD), multi-valve engines (i.e., engines with more than two valves per cylinder), lockup transmissions, engines with variable valve timing, and CVTs. The production fraction for VVT car engines has increased in a similar fashion to the others shown in the figure. This indicates that, in the past, it has taken a decade for a technology to prove itself and attain a production fraction of 40 to 50 percent and as long as another five or ten years to reach maximum market penetration.

Car Technology Penetration
Years After First Significant Use

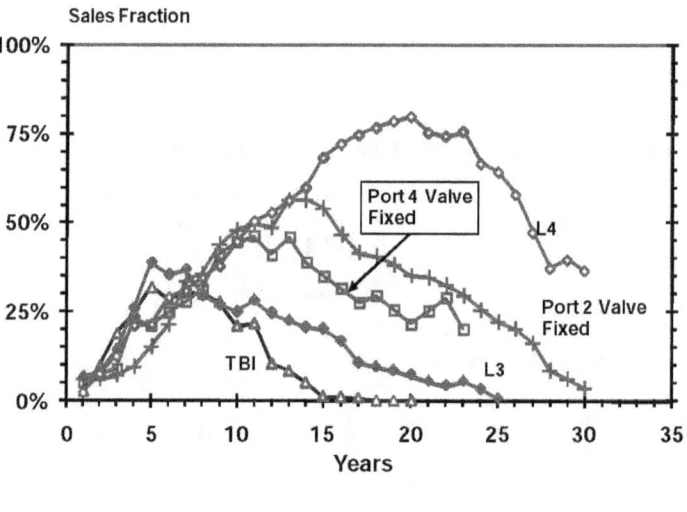

Figure 63

     A similar comparison of five technologies whose production fraction peaked out is shown in Figure 63. This figure shows that, in the past, it has taken a number of years for technologies such as throttle body fuel injection (TBI), lockup 3-speed (L3) and 4-speed (L4) transmissions to reach their maximum production fraction, and, even then, use of these technologies has often continued for a decade or longer. For the limited number of historical cases studied, the time a given technology has taken to attain and then pass a market share of about 40 to 50 percent appears to be one indicator of whether it later attains a stabilized high level of market penetration. L4 transmissions and both two- and four-valve, port injected, fixed valve timing car engines (Port 2V- and 4V- Fixed) now can be classified with technologies such as TBI engines and L3 transmissions which have reached their peak production fractions and, thus, are likely to disappear from the new vehicle fleet.

EPA-420-R-09-014   71   November 2009

Table 25 compares inertia weight, fuel economy ratings, the ratio of highway to city fuel economy, and ton-mpg of the MY2009 hybrid and diesel vehicles with those for the average conventionally powered MY2009 car and truck. All of the hybrid and most of the diesel vehicles in the table have a lower highway/city ratio than the average conventional car or truck.

Table 25

Characteristics of MY 2009 Hybrid and Diesel Vehicles

	IWT	CID	Trans	Lab 55/45 MPG	Adjusted City MPG	Adjusted HWY MPG	Adjusted COMP MPG	HWY/ City Ratio	Ton- MPG
Hybrid Cars									
Prius	3000	91	CVT	65.8	47.7	45.1	46.2	.95	69.3
Altima	3500	152	CVT	46.7	35.1	33.0	33.9	.94	59.3
GS 450H	4500	211	L6	30.8	21.9	25.3	23.8	1.15	53.5
Civic	3000	82	CVT	58.8	40.2	45.3	42.9	1.13	64.4
Camry	4000	144	CVT	45.9	33.4	34.1	33.8	1.02	67.6
Malibu	4000	145	L4	38.6	25.8	34.0	29.9	1.32	59.8
Aura	4000	145	L4	38.6	25.8	34.0	29.9	1.32	59.8
LS 600HL	5500	303	L8	26.9	19.6	21.8	20.8	1.11	57.2
Hybrid Trucks									
Aspen 4WD	6000	348	L4	26.9	19.7	21.7	20.8	1.10	62.3
C15 Sierra 2WD	6000	364	CVT	28.2	21.1	21.7	21.5	1.03	64.4
C15 Silverado 2WD	6000	364	CVT	28.2	21.1	21.7	21.5	1.03	64.4
C1500 Tahoe 2WD	6000	364	CVT	28.2	21.1	21.7	21.5	1.03	64.4
C1500 Yukon 2WD	6000	364	CVT	28.2	21.1	21.7	21.5	1.03	64.4
Escalade 2WD	6000	364	CVT	28.2	21.1	21.7	21.5	1.03	64.4
Escape Hybrid 4WD	4000	140	CVT	37.3	28.7	26.6	27.5	.93	55.0
Escape Hybrid FWD	4000	140	CVT	44.2	34.0	30.7	32.0	.90	64.1
Highlander 4WD	5000	202	CVT	35.2	27.3	25.1	26.0	.92	65.0
K15 Sierra 4WD	6000	364	CVT	28.2	21.1	21.7	21.5	1.03	64.4
K15 Silverado 4WD	6000	364	CVT	28.2	21.1	21.7	21.5	1.03	64.4
K1500 Tahoe 4WD	6000	364	CVT	28.2	21.1	21.7	21.5	1.03	64.4
K1500 Yukon 4WD	6000	364	CVT	28.2	21.1	21.7	21.5	1.03	64.4
Mariner 4WD	4000	140	CVT	37.3	28.7	26.6	27.5	.93	55.0
Mariner FWD	4000	140	CVT	44.2	34.0	30.7	32.0	.90	64.1
Tribute 2WD	4000	140	CVT	44.2	34.0	30.7	32.0	.93	64.1
Tribute 4WD	4000	140	CVT	37.3	28.7	26.6	27.5	.90	55.0
Vue 2-Mode	4500	220	CVT	37.4	26.7	29.6	28.3	1.11	63.6
Vue	4000	145	L4	36.7	24.8	32.2	28.5	1.30	57.0
Diesel Cars									
R320 Bluetec	5500	182	L7	26.3	17.9	23.9	20.9	1.33	57.5
E320 Bluetec	4000	182	L7	34.7	22.8	32.2	27.3	1.41	54.7
Jetta	3500	120	M6	45.5	29.5	40.7	35.0	1.38	61.3
Jetta	3500	120	L6	45.0	29.5	39.8	34.6	1.35	60.5
Jetta Sportwagen	3500	120	M6	45.5	29.5	40.7	35.0	1.38	61.3
Jetta Sportwagen	3500	120	L6	45.0	29.5	39.8	34.6	1.35	60.5
Diesel Trucks									
Touareg	5500	181	L6	26.2	17.5	24.7	21.0	1.41	57.7
Q7	6000	181	L6	24.2	15.9	23.8	19.6	1.50	58.9
GL320 Bluetec	6000	182	L7	24.8	16.9	22.7	19.8	1.34	59.3
Ml320 Bluetec	5000	182	L7	26.2	17.7	24.2	20.9	1.37	52.3
Average Car	3533	167	--	30.9	20.5	28.8	24.5	1.40	43.8
Average Truck	4712	238	--	22.9	15.6	21.4	18.4	1.37	43.5

In addition, there are several cases in the table for which the highway to city ratio is less than 1.0, and these represent cases where a vehicle achieves higher fuel economy in city than in highway driving. This year's diesel cars achieve ton-mpg values that are roughly the same as some of the hybrid cars. For MY2009, the Toyota Prius achieves 69 Ton-mpg, 60 percent higher than that of the average car.

Most of the vehicles in Table 25 have conventionally powered counterparts. Tables 26 and 27 compare the adjusted composite fuel economy and an estimate of annual fuel usage (assuming 15,000 miles per year) for these vehicles with their conventionally powered (baseline) counterparts. The comparisons in both tables are limited to a basis of model name, drive, inertia weight, transmission, and engine size (CID), and for simplicity there is only one listing for "twin" vehicles such as the Escape/Mariner and the Highlander/RX400 H. Differences in the performance attributes of these vehicles complicate making the forward analysis of the fuel economy improvement potential due to hybridization and dieselization. In particular, hybrid vehicles are sometimes reported to have faster 0-to-60 acceleration times than their conventional counterparts, while vehicles equipped with diesel engines have higher low-end torque, but slower 0-to-60 times. In addition, some hybrid vehicles use technologies such as cylinder deactivation and CVT transmissions that are not offered in their counterparts. Given the difficulty in choosing the "right" baseline vehicle, Table 26 includes a comparison for the CVT-equipped Escape Hybrid FWD with baseline data for both manual and automatic transmission versions of this vehicle.

Fuel economy improvements and fuel savings per year for the hybrid vehicles in Table 26 vary considerably from about five percent for the larger, luxury hybrid vehicles to around 40 percent for several others. Similarly, fuel economy improvements for diesels range from 17 to 41 percent, and these vehicles also offer relatively high fuel savings. Nine years after the introduction for sale in the U.S. of the first hybrid vehicle, the MY2000 Honda Insight, hybrid vehicles now account for about two percent of the combined car/truck fleet. In addition, the production fraction for diesels remains at or below 0.5 percent, an order of magnitude smaller than their 5.9 percent production fraction in 1981.

EPA-420-R-09-014

Table 26

Comparison of MY2009 Hybrid Vehicles With Their Conventional Counterparts

Model Name	<----- Hybrid Version ------>				<---- Baseline Version --->					<Improvement>	
	Inertia Weight	CID	Trans	ADJ COMP MPG	Gal Per Year*	Inertia Weight	CID	Trans	ADJ COMP MPG	Gal Per Year*	ADJ COMP MPG	Gal Per Year*
Altima	3500	152	CVT	33.9	443	3500	152	CVT	26.9	559	26%	116
Civic	3000	82	CVT	42.9	349	3000	110	L5	30.6	490	40%	141
						3000	110	M5	30.3	494	41%	145
Camry	4000	144	CVT	33.8	444	3500	144	L5	26.1	574	29%	130
						3500	144	M5	25.8	582	31%	138
Malibu	4000	145	L4	29.9	502	3500	145	L4	25.8	581	16%	79
GS 450H**	4500	211	L6	23.8	631	4000	211	L6	22.4	669	6%	38
LS 600HL**	5500	303	L8	20.8	721	4500	281	L8	20.1	753	4%	32
Aspen 4WD	6000	348	L4	20.8	722	5500	348	L5	16.1	934	29%	212
Escalade 2WD	6000	364	CVT	21.5	699	6000	380	L6	15.4	973	39%	274
Vue (2-Mode)	4500	220	CVT	28.2	530	4000	218	L6	22.0	750	41%	220
Vue	4000	145	L4	28.5	526	4000	145	L4	22.6	663	26%	136
Escape FWD	4000	140	CVT	32.0	468	3500	140	L6	24.0	625	33%	157
						3500	140	M5	25.0	600	28%	132
Escape 4WD	4000	140	CVT	27.5	546	3500	140	L6	22.0	681	25%	136
Highlander 4WD	5000	202	CVT	26.0	577	4500	211	L5	19.7	760	32%	183
C1500 Tahoe 2WD	6000	364	CVT	21.5	699	6000	380	L6	15.4	973	39%	274
K1500 Tahoe 4WD	6000	364	CVT	20.1	745	6000	380	L6	16.0	997	43%	268
C15 Silverado 2WD	6000	364	CVT	21.5	699	5500	380	L6	15.5	968	39%	270
K15 Silverado 4WD	6000	364	CVT	20.1	745	6000	380	L6	16.0	997	43%	268

*Note: Gallons per year calculation is based on all vehicles being driven 15,000 miles.

**Note: Baseline version used for the GS 450H comparison is the GS350. Baseline vehicle used for the LS 600HL comparison is the LS 460L.

EPA-420-R-09-014

Table 27

Comparison of MY2009 Diesel Vehicles With Their Conventional Counterparts

Model Name	<----- Diesel Version ------>					<---- Baseline Version --->					<Improvement>	
	Inertia Weight	CID	Trans	ADJ COMP MPG	Gal Per Year*	Inertia Weight	CID	Trans	ADJ COMP MPG	Gal Per Year*		ADJ COMP MPG	Gal Per Year*
E320 Bluetec**	4000	182	L7	27.3	549	4000	213	L7	20.3	741		35%	192
R320 Bluetec**	5500	182	L7	20.9	717	5500	213	L7	17.0	881		23%	164
Jetta	3500	120	M6	35.0	428	3500	121	M6	24.9	603		41%	175
Jetta	3500	120	L6	34.6	434	3500	121	L6	25.6	587		35%	153
ML320 Bluetec**	5000	182	L7	20.9	717	5000	213	L7	17.2	873		22%	156
GL320 Bluetec**	6000	182	L7	19.8	759	6000	285	L7	15.3	978		29%	218
Touareg	5500	181	L6	21.0	715	5500	219	L6	16.8	893		25%	178
Q7	6000	181	L6	19.6	764	5500	219	L6	16.8	893		17%	129

*Note: Gallons per year calculation is based on all vehicles being driven 15,000 miles.

**Note: Baseline version used for the R320 Bluetec comparison is the R350 4MATIC. Baseline version used for the GL320 Bluetec comparison is the GL450 4MATIC. Baseline version used for the E320 Bluetec comparison is the E350. Baseline version used for the ML320 Bluetec comparison is the ML350 4MATIC.

VII. Marketing Groups and Fuel Economy

In its century of evolution, the automotive industry existed first as small, individual companies that relatively quickly went out of business or grew into larger corporations. Prior to the 1970s, the historic term "manufacturer" usually meant an automobile company that manufactured and sold vehicles in its own country and perhaps exported vehicles to a few other countries. Over the years, the nature of the automotive industry has changed substantially, and it has evolved into one in which global consolidations and alliances among heretofore independent manufacturers have become the norm, rather than the exception.

Early reports in this series examined fuel economy and technology trends for the "Domestic" and "Import" vehicle categories which are part of the corporate average fuel economy program. Over time, this classification approach evolved into a market segment approach in which cars were apportioned to a "Domestic," "European," and "Asian" category, with trucks classified as "Domestic" or "Imported." As the automotive industry has become more transnational in nature, this type of vehicle classification has become less useful. In the most recent reports in this series, trends by groups of manufacturers have been used to reflect the transnational and transregional nature of the automobile industry.

There are 33 individual manufacturers in the 2009 CO_2 and fuel economy trends database. To reflect the transition to an industry in which there are a smaller number of independent companies, these 33 individual manufacturers have been divided into nine major marketing group segments, and a tenth catch-all group ("Others") that contains smaller manufacturers not assigned to one of the nine major marketing groups.

These nine major marketing groups are:

1. The General Motors Group includes GM, Daewoo, Saab, and Isuzu;

2. The Ford Motor Group includes Ford, Volvo, Roush, and Saleen;

3. The Chrysler Group includes only Chrysler;

4. The Toyota Group includes only Toyota;

5. The Honda Group includes only Honda;

6. The Nissan Group includes only Nissan;

7. The Hyundai-Kia (HK) Group includes Hyundai and Kia;

8. The VW Group includes Volkswagen, Audi, Bentley, and Lamborghini; and

9. The BMW group includes BMW and Phantom.

Taken together, the nine major marketing groups comprise over 95 percent of the MY2009 new vehicle market in the U.S. It is expected that these marketing groups will continue to evolve and perhaps expand, or possibly contract as further changes in the automotive industry occur. The changes in the marketing group definitions for this report are that Mazda, Rover, and Jaguar are moved out of the Ford marketing group.

Tables 28 and 29 list the 33 individual manufacturers which are included in EPA's 2009 database, and the marketing group to which they are assigned for this report. Table 28 shows the projected MY2009 laboratory 55/45 fuel economy values for cars only, trucks only, and cars and trucks combined, along with the

EPA-420-R-09-014   76   November 2009

truck market share, for each of the 33 individual manufacturers. Table 29 shows the same information, but with projected MY2009 adjusted composite fuel economy values instead.

Tables 30 and 31 provide fuel economy data for the nine marketing groups, with the former providing laboratory 55/45 fuel economy data, and the latter including adjusted composite fuel economy data. The bottom two rows in each table give the overall average MY2009 fuel economy value, as well as the truck market share, for each marketing group. It can be seen that the Honda, Hyundai-Kia, and Toyota marketing groups have the highest projected MY2009 fuel economy values. Chrysler has the lowest projected MY2009 fuel economy value. Tables 30 and 31 also show the average marketing group fuel economies by vehicle type and size. For example, Table 30 shows that Hyundai-Kia has the highest projected MY2009 laboratory 55/45 fuel economy value for the small car class. Different marketing groups are leaders in other vehicle classes as defined by this report.

Table 32 combines MY2008 vehicle footprint and fuel economy data by marketing group. MY2008 is shown here for two reasons: it is the only year for which we have footprint data, and it is the most recent year for which we have essentially final fuel economy data based on actual production as reported in the end-of-year CAFE reports. For MY2008, Volkswagen had the lowest fleetwide footprint, while Honda had the highest fleetwide fuel economy, followed closely by Hyundai-Kia. General Motors had the highest footprint, with Chrysler having the lowest fleetwide adjusted fuel economy and Ford close behind.

Figures 64 through 72 compare, on a 3-year moving average basis, the percent truck and laboratory 55/45 fuel economy for cars, trucks, and both cars and trucks for the nine marketing groups. More information stratified by marketing group can be found in the Appendices L through O.

It is important to note when a marketing group definition is changed to reflect a change in the industry's financial arrangements, EPA makes the same adjustment in marketing group composition in the historical database that is used for Figures 64 through 72 and in Appendices L through O, as well. This maintains a consistent marketing group definition over time, which allows a better identification of long-term trends. On the other hand, this also means that the database does not necessarily reflect actual financial arrangements in the past. For example, the 2009 database no longer accounts for the fact that Chrysler was combined with Daimler for several years.

EPA-420-R-09-014   77   November 2009

Table 28

Model Year 2009 Laboratory 55/45 Fuel Economy by Manufacturer

Manufacturer	Marketing Group	Cars	Trucks	Both	Percent Truck
General Motors	General Motors	29.0	21.8	24.5	56%
Toyota	Toyota	35.2	24.2	29.4	43%
Chrysler	Chrysler	27.4	22.2	23.2	77%
Honda	Honda	33.7	25.5	29.7	42%
Nissan	Nissan	32.8	22.5	27.2	45%
Ford	Ford	28.7	23.5	25.7	53%
Hyundai	Hyundai-Kia	32.2	25.7	30.1	27%
Kia	Hyundai-Kia	33.4	24.0	28.0	49%
Volkswagen	Volkswagen	31.8	24.6	29.6	26%
BMW	BMW	28.0	22.3	26.9	15%
Daimler AG	Other	25.2	20.5	24.0	22%
Subaru	Other	28.5	26.6	27.6	45%
Mazda	Other	30.0	23.4	27.6	30%
Mitsubishi	Other	29.3	25.9	28.2	29%
Audi	Volkswagen	28.5	21.9	26.6	23%
GM Daewoo	General Motors	37.8		37.8	0%
Suzuki	Other	33.1	25.4	29.7	39%
Volvo	Ford	25.8	20.7	24.1	29%
Rover	Other		19.3	19.3	100%
Porsche	Other	27.4	20.0	22.6	58%
Jaguar	Other	24.2		24.2	0%
Saab	General Motors	26.5	20.4	25.6	12%
Maserati	Other	18.3		18.3	0%
Bentley	Volkswagen	15.7		15.7	0%
Isuzu	General Motors		20.7	20.7	100%
Ferrari	Other	16.4		16.4	0%
Aston Martin	Other	18.2		18.2	0%
Roush	Ford	21.3		21.3	0%
Lamborghini	Volkswagen	16.1		16.1	0%
Phantom	BMW	17.3		17.3	0%
Lotus	Other	30.0		30.0	0%
Saleen	Ford	17.4		17.4	0%
Spyker	Other	19.3		19.3	0%
Fleet		30.9	22.9	26.4	49%

Table 29

Model Year 2009 Adjusted Composite Fuel Economy by Manufacturer

Manufacturer	Marketing Group	<-- FUEL ECONOMY -->			Percent Truck
		Cars	Trucks	Both	
General Motors	General Motors	23.3	17.6	19.7	56%
Toyota	Toyota	27.4	19.3	23.2	43%
Chrysler	Chrysler	21.9	17.9	18.7	77%
Honda	Honda	26.6	20.4	23.6	42%
Nissan	Nissan	25.8	18.0	21.6	45%
Ford	Ford	22.9	18.8	20.6	53%
Hyundai	Hyundai-Kia	25.5	20.4	23.9	27%
Kia	Hyundai-Kia	26.3	19.2	22.3	49%
Volkswagen	Volkswagen	25.1	19.7	23.5	26%
BMW	BMW	22.5	17.9	21.7	15%
Daimler AG	Other	20.3	16.5	19.3	22%
Subaru	Other	22.5	21.1	21.9	45%
Mazda	Other	23.8	18.7	22.0	30%
Mitsubishi	Other	23.2	20.6	22.4	29%
Audi	Volkswagen	22.6	17.5	21.2	23%
GM Daewoo	General Motors	29.5		29.5	0%
Suzuki	Other	26.0	20.2	23.4	39%
Volvo	Ford	20.8	16.8	19.4	29%
Rover	Other		15.7	15.7	100%
Porsche	Other	22.0	16.3	18.3	58%
Jaguar	Other	19.7		19.7	0%
Saab	General Motors	21.5	16.5	20.7	12%
Maserati	Other	15.0		15.0	0%
Bentley	Volkswagen	13.1		13.1	0%
Isuzu	General Motors		16.7	16.7	100%
Ferrari	Other	13.5		13.5	0%
Aston Martin	Other	15.0		15.0	0%
Roush	Ford	17.2		17.2	0%
Lamborghini	Volkswagen	13.3		13.3	0%
Phantom	BMW	14.2		14.2	0%
Lotus	Other	23.6		23.6	0%
Saleen	Ford	14.4		14.4	0%
Spyker	Other	15.6		15.6	0%
Fleet		24.5	18.4	21.1	49%

EPA-420-R-09-014

Table 30

Model Year 2009 Laboratory 55/45 Fuel Economy by Marketing Group

VEHICLE TYPE/SIZE	GM	Toyota	Chrysler	Honda	Nissan	Ford	HK	VW	BMW	All
Cars										
Small	32.2	37.5	24.8	36.8	28.6	31.9	39.2	31.3	29.0	32.6
Midsize	29.9	34.4	31.1	26.5	33.8	27.9	35.3	22.9	25.9	31.7
Large	26.1	29.6	25.8	31.2	23.7	24.4	30.5	23.7	21.6	27.5
All	29.1	35.5	27.4	33.3	32.7	28.6	32.6	30.2	28.0	30.9
Wagons										
Small	32.0	32.4	27.5	39.9	37.9	29.1	34.5	34.3	26.7	32.6
Midsize	26.4					23.6	28.0	28.4	24.6	27.7
Large										21.8
All	31.9	32.4	27.5	39.9	37.9	23.7	29.7	33.5	25.7	31.4
All Cars										
Small	32.1	36.4	25.7	37.3	29.9	31.9	38.4	31.6	29.0	32.6
Midsize	29.9	34.4	31.1	26.5	33.8	27.3	34.4	23.5	25.9	31.5
Large	26.1	29.6	25.8	31.2	23.7	24.4	30.5	23.7	21.6	27.5
All	29.3	35.2	27.4	33.7	32.8	28.4	32.4	30.5	28.0	30.9
Vans										
Small										
Midsize		26.2	24.6	25.4	24.5	24.2	23.8	24.1		24.9
Large	19.7									19.7
All	19.7	26.2	24.6	25.4	24.5	24.2	23.8	24.1		24.6
SUVs										
Small			21.4							23.5
Midsize	27.9	26.0	23.0	25.8	25.9	26.4	25.5	26.6		25.3
Large	22.1	19.1	22.8		22.7	22.7	23.3	21.6	22.3	22.2
All	22.5	25.1	22.7	25.8	23.7	24.6	25.1	23.8	22.3	23.6
Pickups										
Small										
Midsize	24.9	24.4				25.1				24.6
Large	20.5	19.3	19.8	22.0	19.6	20.4				20.1
All	20.6	22.3	19.8	22.0	19.6	21.4				20.8
Trucks										
Small			21.4							23.5
Midsize	27.3	25.6	23.6	25.7	25.8	26.1	25.1	25.1		25.2
Large	21.5	19.2	21.2	22.0	21.5	21.6	23.3	21.6	22.3	21.3
All	21.8	24.2	22.2	25.5	22.5	23.4	24.9	23.9	22.3	22.9
Fleet										
All	24.7	29.4	23.2	29.7	27.2	25.6	29.4	28.6	26.9	26.4
Truck %	55%	43%	77%	42%	45%	51%	34%	25%	15%	49%

EPA-420-R-09-014

Table 31

Model Year 2009 Adjusted Composite Fuel Economy by Marketing Group

VEHICLE TYPE/SIZE	GM	Toyota	Chrysler	Honda	Nissan	Ford	HK	VW	BMW	All
Cars										
Small	25.6	29.0	20.1	28.9	22.7	25.3	30.4	24.8	23.2	25.7
Midsize	24.0	27.0	24.5	21.3	26.5	22.3	27.6	18.4	21.0	25.1
Large	21.1	23.7	20.8	25.0	19.1	19.8	24.4	19.1	17.6	22.2
All	23.3	27.7	21.9	26.4	25.7	22.8	25.8	24.0	22.5	24.5
Wagons										
Small	25.2	25.3	21.7	30.5	28.7	23.3	26.9	26.9	21.5	25.5
Midsize	21.5					19.1	22.3	22.8	19.9	22.0
Large										17.4
All	25.2	25.3	21.7	30.5	28.7	19.2	23.5	26.3	20.7	24.7
All Cars										
Small	25.5	28.2	20.7	29.1	23.6	25.3	29.8	25.0	23.2	25.7
Midsize	24.0	27.0	24.5	21.3	26.5	21.9	27.0	18.9	20.9	24.9
Large	21.1	23.7	20.8	25.0	19.1	19.8	24.4	19.1	17.6	22.1
All	23.5	27.4	21.9	26.6	25.8	22.7	25.7	24.2	22.5	24.5
Vans										
Small										
Midsize		20.9	19.8	20.5	19.7	19.5	19.2	19.4		20.1
Large	15.8									15.8
All	15.8	20.9	19.8	20.5	19.7	19.5	19.2	19.4		19.8
SUVs										
Small			17.0							18.7
Midsize	22.2	20.6	18.3	20.5	20.6	21.1	20.2	21.2		20.2
Large	17.9	15.5	18.4		18.2	18.3	18.8	17.3	17.9	17.9
All	18.2	19.9	18.2	20.5	18.9	19.7	20.0	19.1	17.9	19.0
Pickups										
Small										
Midsize	19.9	19.3				19.9				19.5
Large	16.5	15.5	16.0	17.6	15.8	16.4				16.2
All	16.6	17.8	16.0	17.6	15.8	17.2				16.7
Trucks										
Small			17.0							18.7
Midsize	21.7	20.3	18.9	20.5	20.5	20.7	20.0	20.2		20.1
Large	17.3	15.5	17.1	17.6	17.3	17.4	18.8	17.3	17.9	17.2
All	17.6	19.3	17.9	20.4	18.0	18.7	19.9	19.2	17.9	18.4
Fleet										
All	19.9	23.2	18.7	23.6	21.6	20.5	23.4	22.8	21.6	21.1
Truck %	55%	43%	77%	42%	45%	51%	34%	25%	15%	49%

EPA-420-R-09-014   81   November 2009

Table 32

MY2008 Footprint and Fuel Economy by Marketing Group

Marketing Group	Vehicle Type	Footprint SQFT	Lab 55/45 MPG	Adjusted Composite MPG
General Motors	Cars	46.2	28.6	23.0
General Motors	Trucks	56.6	21.6	17.4
General Motors	All	51.7	24.4	19.7
Toyota	Cars	44.1	36.0	28.1
Toyota	Trucks	52.8	23.9	19.0
Toyota	All	48.3	29.0	22.8
Chrysler	Cars	47.4	27.8	22.2
Chrysler	Trucks	49.9	22.4	18.0
Chrysler	All	48.9	24.2	19.3
Honda	Cars	44.7	34.3	27.1
Honda	Trucks	48.4	25.5	20.3
Honda	All	46.2	30.1	23.9
Nissan	Cars	45.4	32.2	25.3
Nissan	Trucks	52.8	22.0	17.7
Nissan	All	48.1	27.6	21.9
Ford	Cars	46.4	27.9	22.4
Ford	Trucks	53.5	22.2	17.8
Ford	All	50.8	24.2	19.4
Hyundai-Kia	Cars	44.5	33.6	26.5
Hyundai-Kia	Trucks	48.2	24.9	19.9
Hyundai-Kia	All	45.8	30.0	23.7
Volkswagen	Cars	43.6	28.9	23.1
Volkswagen	Trucks	52.8	20.2	16.3
Volkswagen	All	44.4	27.9	22.3
BMW	Cars	45.4	27.2	21.9
BMW	Trucks	50.0	22.9	18.5
BMW	All	46.2	26.3	21.2
Fleet	Cars	45.4	30.5	24.3
Fleet	Trucks	52.9	22.7	18.2
Fleet	All	49.0	26.3	21.0

EPA-420-R-09-014

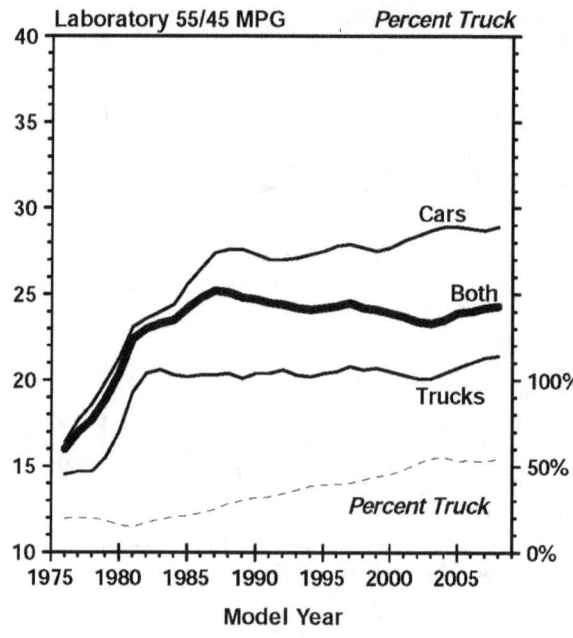

**GM Marketing Group
Fuel Economy by Model Year
(Three Year Moving Average)**

Figure 64

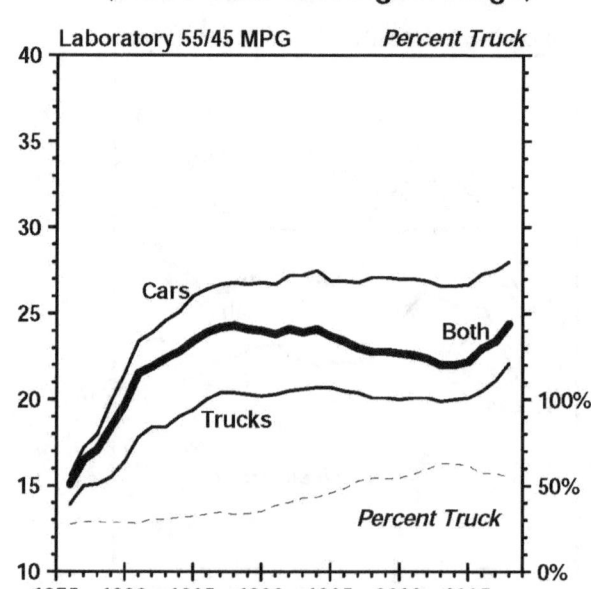

**Ford Marketing Group
Fuel Economy by Model Year
(Three Year Moving Average)**

Figure 65

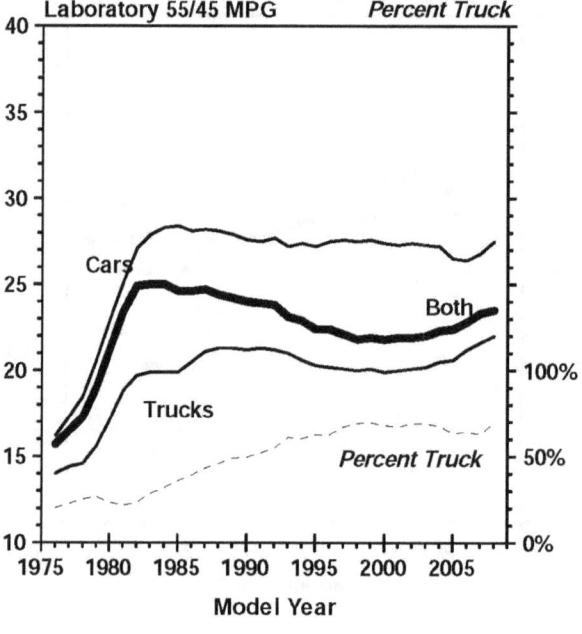

**Chrysler Marketing Group
Fuel Economy by Model Year
(Three Year Moving Average)**

Figure 66

EPA-420-R-09-014   November 2009

Toyota Marketing Group
Fuel Economy by Model Year
(Three Year Moving Average)

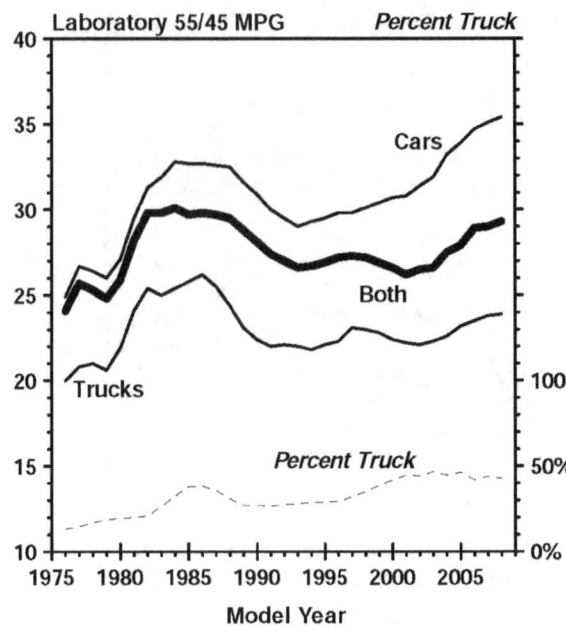

Figure 67

Honda Marketing Group
Fuel Economy by Model Year
(Three Year Moving Average)

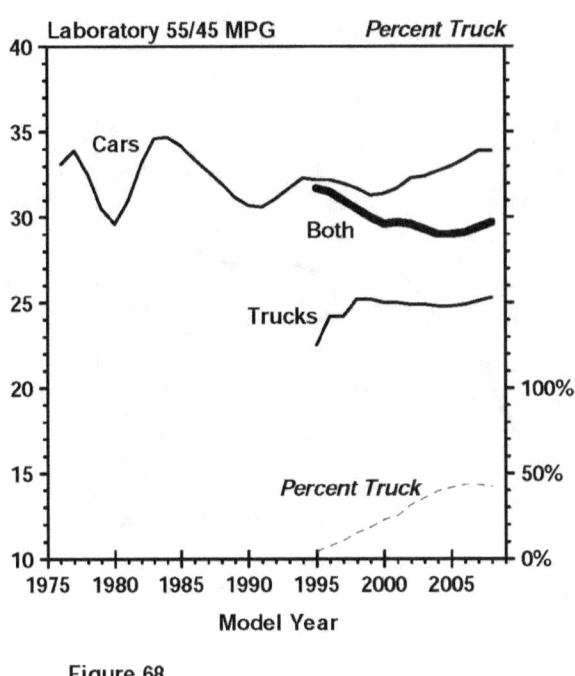

Figure 68

Nissan Marketing Group
Fuel Economy by Model Year
(Three Year Moving Average)

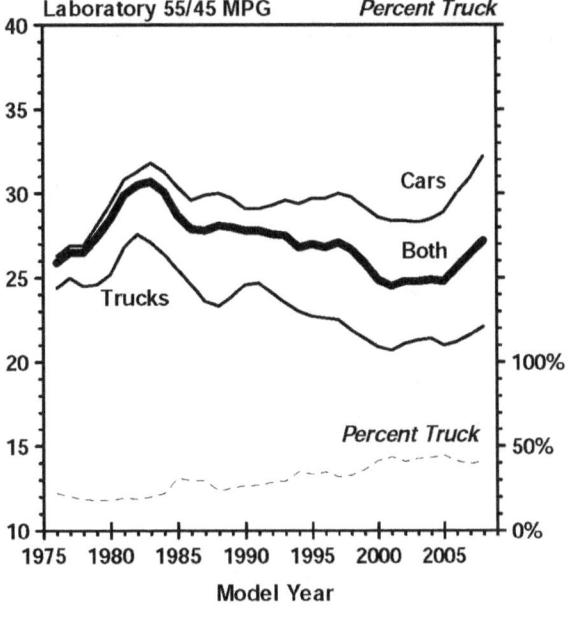

Figure 69

EPA-420-R-09-014　　　　　　84　　　　　　November 2009

Hyundai-Kia Marketing Group
Fuel Economy by Model Year
(Three Year Moving Average)

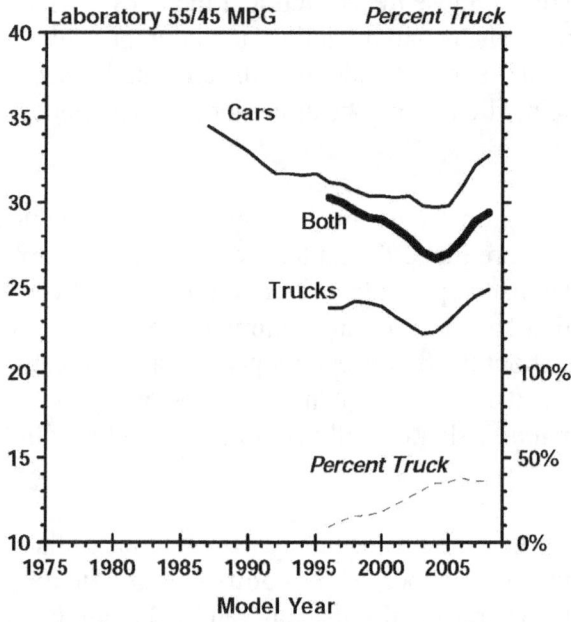

Figure 70

VW Marketing Group
Fuel Economy by Model Year
(Three Year Moving Average)

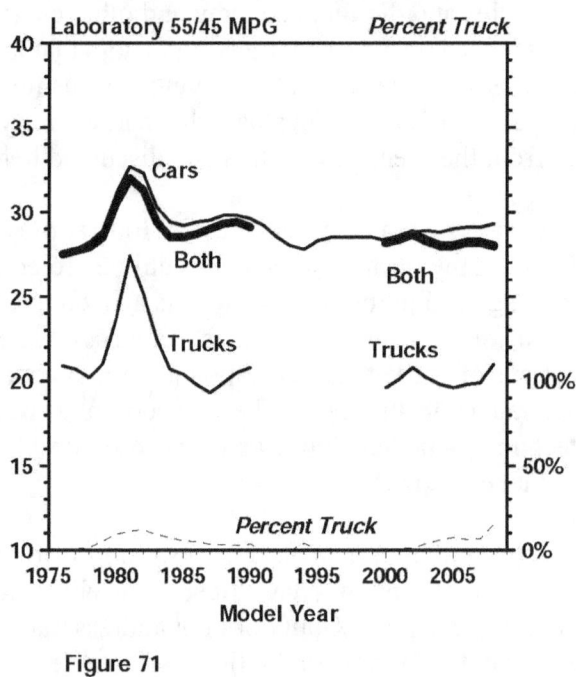

Figure 71

BMW Marketing Group
Fuel Economy by Model Year
(Three Year Moving Average)

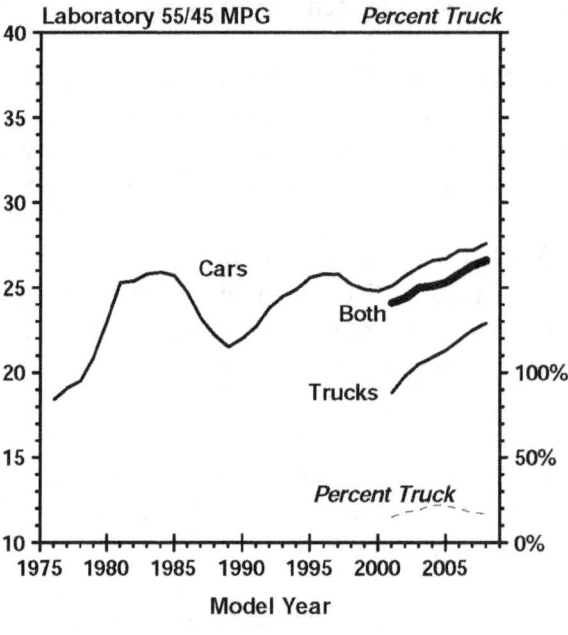

Figure 72

EPA-420-R-09-014   85   November 2009

VIII. <u>Characteristics of Fleets Comprised of Existing Fuel-Efficient Vehicles</u>

This section is limited to a discussion of hypothetical fleets of vehicles comprised of existing fuel-efficient vehicles and the fuel economy and other characteristics of those fleets. While it includes a discussion of some of the technical and engineering factors that affect fleet fuel economy, it does not attempt to evaluate either the benefits or the costs of achieving various fuel economy levels. In addition, the analysis presented here also does not attempt to evaluate the marketability or the public acceptance of any of the hypothetical fleets that result from the scenarios studied and discussed below.

There are several different ways to look at the potential for improved fuel economy from the light-duty vehicle fleet. Many of these approaches utilize projections of more fuel efficient technologies that are not currently being used in the fleet today. As an example, a fleet made up of a large fraction of fuel cell vehicles could be considered. Such projections can be associated with a good deal of uncertainty, since uncertainty in the projections of market share compound with uncertainties about the fuel economy performance of yet uncommercialized technology. These uncertainties can be thought of as a combination of technical risk, i.e., can the technology be developed and mass produced?, and market risk, i.e., will people buy vehicles with the improved fuel economy?

One general approach used in this report is to consider only the fuel economy performance of those technologies which exist in today's fleet. This eliminates uncertainty about the feasibility and production readiness of the technology, but does not address market risk. Therefore, the analysis can be thought of as the fuel economy potential now in the fleet, with no new technologies added, if the higher mpg choices available were to be selected by a much higher percentage of consumers.

As was shown in Figures 3 and 4, there is a wide distribution of fuel economy. Because of the interest in the high end of this spectrum, this portion of the database was examined in more detail using three "best in class" (BIC) analysis techniques. This type of technique is not new, and in fact was one of the methods used to investigate future fleet fuel economy capability when the original fuel economy standards were set.

In any group or class of vehicles there will be a distribution of fuel economy performance, and the "best in class" method relies on that fact. The analysis involves dividing the fleet of vehicles into classes, selecting a set of representative high mpg "role model" vehicles from each class, and then calculating the average characteristics of the resultant fleet using the same relative production proportions as in the baseline fleet.

One potential problem with a BIC analysis is that the high mpg cars used in the analysis may be unusual in some way - so unusual that the hypothetical BIC fleet may be deficient in some other attributes considered desirable by vehicle buyers. Because the BIC analysis is also sensitive to the selection of the best vehicles, three different procedures were used to select the role models.

Two of these selection procedures use the EPA car size classes (which for cars are the same as those used for the EPA/DOE *Fuel Economy Guide*) and the truck type/size classes described previously in this report. The third best-in-class role model selection procedure is based on using the vehicle inertia weight classes used for EPA's vehicle testing and certification programs.

The advantage of using and analyzing data from the best-in-size class methods is that if the production proportions of each class are held constant, the production distribution of the resultant fleet by *vehicle type and size* does not change. This means that the size of the average vehicle does not change a lot, but there can be some fluctuation in interior volume for cars because of the distribution of interior volume within a car class. Similarly, another advantage of using the inertia weight classes to determine the role models is, if the

EPA-420-R-09-014
November 2009

production proportions in each inertia weight class are held constant, the production distribution of the resultant fleet by *weight* does not change, and in this case, the average weight remains the same.

One way of performing a best-in-class analysis is to use as role models the four nameplates with the highest fuel economy in each size class. (See Tables Q-1 and Q-2 in Appendix Q.) Under this procedure, all vehicles in a class with the same nameplate are included as role models regardless of vehicle configuration. Each role model nameplate from each class was assigned the same production weighting factor, but the original production weighting distribution for different vehicle configurations within a given nameplate (e.g., transmission type, engine size, and/or drive type) was retained. The resulting values were used to recalculate the fleet average values using the same relative proportions in each of the size classes that constitute the fleet. In cases where two identical vehicles differ by only one characteristic but have slightly different nameplates (such as the two-wheel drive Chevrolet C1500 and the four-wheel drive Chevrolet K1500 pickups), both are considered to be different nameplates. Conversely, in the cases where there are technically identical vehicles with different nameplates, only one representative vehicle nameplate was considered in the BIC analysis.

The second best-in-class role model selection procedure involves selecting as role models the best dozen vehicles in each size class with each vehicle configuration (some of which may have the same nameplate) considered separately. Tables Q-3 and Q-4 in Appendix Q give listings of the representative vehicles used in this method. As with the previous procedure, in cases where technically identical vehicle configurations have different nameplates, only one representative vehicle was considered. Under this best-in-class method, the production data for each role model vehicle in each class was assigned the same value, and the resulting values were used to re-calculate the fleet values again using the same relative proportions in each of the size classes that constitute the fleet.

The third best-in-class procedure involves selecting as role models the best dozen vehicles in each weight class. As with the previous method, each vehicle configuration was considered separately. (See Tables Q-5 and Q-6 in Appendix Q for a listing of the vehicles used in this analysis.) It should be noted that some of the weight classes have less than a dozen representative vehicles. In addition, as in the previous two best-in-class methods, where technically identical vehicle configurations with different nameplates exist, only one representative vehicle was included. As with the two best-in-size class methods, the production data for each role model vehicle in each class was assigned the same value, and the resulting values were used to recalculate the fleet values again using the same relative proportions in each of the size classes that constitute the fleet.

Tables 33 to 35 compare, for cars, trucks, and both cars and trucks, respectively, the results of the best-in-class analysis with actual average data for model year 2009. As discussed earlier, for the size class scenarios, the percentage of vehicles that are small, midsize, or large are the same as for the baseline fleet, and in the weight class scenarios, the average weight of the BIC data sets is the same as the actual one.

In general, the vehicles used for the BIC analysis have less powerful engines, have slower 0-to-60 acceleration times, and are more likely to be equipped with front wheel drive, VVT, CVTs, and hybrid powertrains than the entire fleet as a whole.

Depending on the BIC scenario chosen, MY2009 cars could have achieved from 18 to 27 percent better fuel economy than they did. Similarly, for trucks the potential fuel economy improvement ranges from 13 to 27 percent better fuel economy, and the combined car and truck fleet could have been 15 to 27 percent better.

The best-in-class analyses can be thought of as the mpg potential now in the fleet with no new technologies added if the higher mpg choices available were selected. As such, the best-in-class analyses provide a useful reference point reflecting the variation in fuel economy levels that results in large part from consumer preferences as opposed to technological availability.

EPA-420-R-09-014   87   November 2009

Table 33

Best in Class Results 2009 Cars

Vehicle Characteristic	Selection Basis	Actual Data	Size Class	Size Class	Weight Class
	Selection Criteria	All Cars	Best 4 Nameplates	Best 12 Vehicles	Best 12 Vehicles
Fuel Economy	Lab. 55/45	30.9	39.2	36.9	36.6
	Adjusted City	20.5	26.5	24.7	24.6
	Adjusted Highway	28.8	33.8	32.7	32.2
	Adjusted Composite	24.5	30.2	28.7	28.4
Vehicle Size	Weight (lb.)	3533	3394	3246	3533
	Volume (Cu. Ft)	111	109	109	104
Engine	CID	167	134	130	130
	HP	198	159	153	169
	HP/CID	1.20	1.20	1.18	1.32
	HP/WT	.055	.046	.047	.047
	Percent Multivalve	90%	89%	95%	98%
	Percent Variable Valve	74%	85%	82%	64%
	Percent Diesel	0.8%	6.6%	2.2%	13.5%
Performance	0-60 Time (Sec.)	9.5	9.6	10.3	9.8
	Top Speed	137	126	125	128
	Ton-MPG	43.8	52.6	47.2	50.8
	Cu. Ft. Mpg	2786	3416	3199	3041
	Cu. Ft. Ton-MPG	4858	5775	5182	5311
Drive	Front	79%	98%	94%	79%
	Rear	14%	2%	5%	7%
	4WD	7%	1%	1%	14%
Transmission	Manual	9%	12%	35%	34%
	Lockup	79%	57%	44%	39%
	CVT	11%	30%	21%	26%
Hybrid Vehicle		2.7%	36.7%	12.2%	12.1%

EPA-420-R-09-014   November 2009

Table 34

Best in Class Results 2009 Trucks

Vehicle Characteristic	Selection Basis	Actual Data	Size Class	Size Class	Weight Class
	Selection Criteria	All Trucks	Best 4 Nameplates	Best 12 Vehicles	Best 12 Vehicles
Fuel Economy	Lab. 55/45	22.9	29.1	27.8	25.9
	Adjusted City	15.6	20.9	19.3	17.7
	Adjusted Highway	21.4	23.8	24.1	23.4
	Adjusted Composite	18.4	22.5	21.8	20.5
Vehicle Size	Weight (lb.)	4712	4847	4360	4712
Engine	CID	238	250	205	219
	HP	253	253	222	243
	HP/CID	1.08	1.04	1.11	1.13
	HP/WT	.053	.052	.051	.051
	Percent Multivalve	66%	58%	74%	80%
	Percent Variable Valve	56%	73%	74%	65%
	Percent Diesel	0.1%	<0.1%	5.3%	4.2%
Performance	0-60 Time (Sec.)	9.6	8.3	9.3	9.5
	Top Speed	142	140	136	139
	Ton-MPG	43.5	55.4	48.1	48.7
Drive	Front	29%	33%	34%	37%
	Rear	23%	28%	24%	21%
	4WD	48%	40%	42%	43%
Transmission	Manual	2%	3%	18%	7%
	Lockup	93%	36%	50%	73%
	CVT	5%	62%	33%	19%
Hybrid Vehicle		0.9%	65.8%	24.6%	13.3%

EPA-420-R-09-014

November 2009

Table 35

Best in Class Results 2009 Light Duty Vehicles

Vehicle Characteristic	Selection Basis	Actual Data	Size Class	Size Class	Weight Class
	Selection Criteria	All Vehicles	Best 4 Nameplates	Best 12 Vehicles	Best 12 Vehicles
Fuel Economy	Lab. 55/45	26.4	33.5	31.8	30.4
	Adjusted City	17.8	23.5	21.8	20.6
	Adjusted Highway	24.6	28.0	28.1	27.2
	Adjusted Composite	21.1	25.9	24.9	23.9
Vehicle Size	Weight (lb.)	4107	4102	3789	4107
Engine	CID	202	191	167	173
	HP	225	205	187	205
	HP/CID	1.14	1.12	1.15	1.23
	HP/WT	.054	.049	.049	.049
	Percent Multivalve	79%	74%	84%	87%
	Percent Variable Valve	65%	79%	78%	64%
	Percent Diesel	0.5%	3.4%	4.9%	9.0%
Performance	0-60 Time (Sec.)	9.5	9.0	9.8	9.6
	Top Speed	139	133	130	133
	Ton-MPG	43.6	54.0	47.7	49.8
Drive	Front	55%	66%	64%	59%
	Rear	19%	14%	15%	13%
	4WD	27%	20%	21%	28%
Transmission	Manual	6%	7%	25%	20%
	Lockup	86%	47%	47%	56%
	CVT	8%	46%	27%	23%
Hybrid Vehicle		1.8%	50.9%	18.2%	12.7%

Another general approach for determining potential fuel economy improvement is to study the effects on fuel economy caused by the changes that have occurred in the distributions of vehicle weight and size. This technique involves preserving the average characteristics of vehicles within each size or weight strata in today's fleet, but re-mixing the production distributions to match those of a baseline year and then calculating the fleet wide averages for those characteristics using the re-mixed production data. The production distribution of the resultant fleet is by *vehicle type and size,* thus it is forced to be the same as that for the base year. As with the best in car size class technique, there can be some fluctuation in average interior volume for cars because of the distribution of interior volume within a car class. Similarly, if the production proportions in each inertia weight class are held the same as the base year's, the production distribution of the resultant fleet by *weight* remains the same as that for the base year change, and the recalculated average weight is the same as the base year's.

It is important to note that, for Tables 36 and 37 below, both hybrid and diesel vehicles were excluded so that only vehicles with conventional powertrains were considered. Accordingly, the data in the rows for actual 2009, 1981, and 1988 typically differ slightly from data reported elsewhere in this report.

Table 36 compares weight, interior volume, engine CID and HP, estimated 0-to-60 time and laboratory fuel economy for conventionally powered MY2009 cars as calculated from the projected 2009 production distribution and then recalculated using the size and weight distributions from MY1981 and MY1988. The base years of 1981 and 1988 were chosen because 1981 was the year with the lowest average weight and horsepower levels, and 1988 was the year with the highest LAB fuel economy. This table includes the actual 1981 and 1988 fleet averages as a point of reference. In both of the weight distribution cases, the fuel economy of the re-mixed MY2009 fleet would have been higher than actually is: 10 percent if the 1981 weight distribution is used, 14 percent if the 1988 weight distribution is used. For both re-mixed weight cases, interior volume and horsepower are substantially lower. Using the MY1981 and MY1988 size mix distributions result in a much smaller change of a two percent increase in car fuel economy.

Table 36

Characteristics of MY 2009 Cars

Calculated From:	Inertia Weight	Interior Volume	Engine CID	HP	0 to 60 Time	Lab 55/45 MPG
2009 Actual Distribution	3538	111	168	200	9.5	30.5
1981 Weight Distribution	3043	98	135	172	9.6	33.4
1988 Weight Distribution	3047	103	128	156	10.0	34.9
1981 Size Distribution	3468	107	160	194	9.6	31.0
1988 Size Distribution	3447	108	159	191	9.6	31.1
Reference: 1981 Actual	3043	106	178	99	14.1	24.9
Reference: 1988 Actual	3047	107	160	116	12.8	28.6
Percent Change:						
2009 Actual Distribution	0%	0%	0%	0%	0%	0%
1981 Weight Distribution	-14%	-12%	-20%	-14%	1%	10%
1988 Weight Distribution	-14%	-7%	-24%	-22%	5%	14%
1981 Size Distribution	-2%	-4%	-5%	-3%	1%	2%
1988 Size Distribution	-3%	-3%	-5%	-5%	1%	2%
Reference: 1981 Actual	-14%	-5%	6%	-51%	48%	-18%
Reference: 1988 Actual	-14%	-4%	-5%	-42%	35%	-6%

EPA-420-R-09-014   91   November 2009

Table 37 shows similar data for trucks, and as with the car class cases using either the 1981 or the 1988 production distribution by weight class, results in higher recalculated fuel economy than using the corresponding size class production distribution. Figures 73 to 76 compare actual fuel economy for all model years from 1975 to 2007 with what it would have been had the distributions of weight or size been the same as 1981 or 1988. For both cars and trucks, using either the 1981 or 1988 weight class distribution, results in significantly higher fuel economy improvements than the similar size class cases.

Table 37

Characteristics of MY 2009 Trucks

	Inertia Weight	Engine CID	HP	0 to 60 Time	Lab 55/45 MPG
Calculated From:					
2009 Actual Distribution	4709	239	253	9.6	22.9
1981 Weight Distribution	3841	173	201	9.8	28.2
1988 Weight Distribution	3838	174	195	10.1	28.1
1981 Size Distribution	4532	248	252	9.7	22.7
1988 Size Distribution	4392	227	229	10.0	23.6
Reference: 1981 Actual	3841	252	121	14.4	19.7
Reference: 1988 Actual	3838	227	141	12.9	21.2
Percent Change:					
2009 Actual Distribution	0%	0%	0%	0%	0%
1981 Weight Distribution	-18%	-28%	-21%	2%	23%
1988 Weight Distribution	-18%	-27%	-23%	5%	23%
1981 Size Distribution	-4%	4%	0%	1%	-1%
1988 Size Distribution	-7%	-5%	-9%	4%	3%
Reference: 1981 Actual	-18%	5%	-52%	50%	-14%
Reference: 1988 Actual	-18%	-5%	-44%	34%	-7%

EPA-420-R-09-014

Effect of Weight and Size On Car Fuel Economy

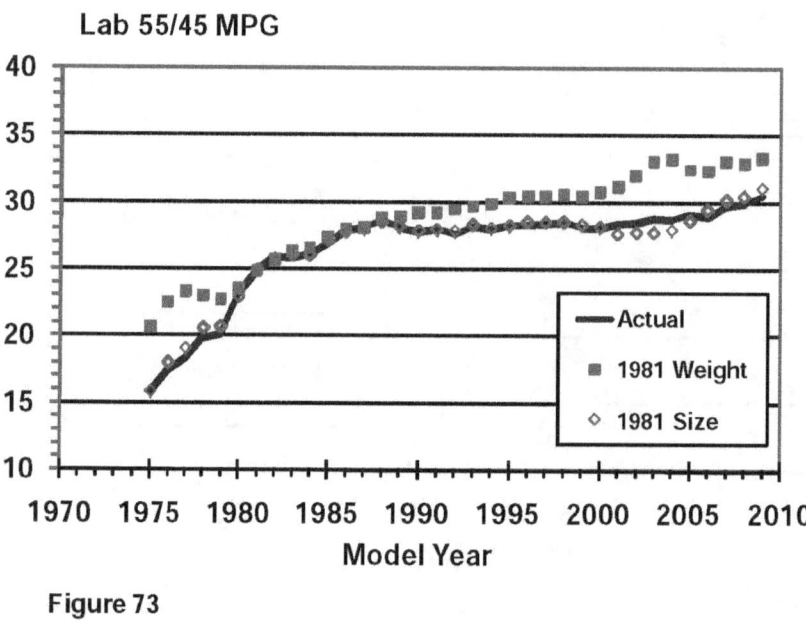

Figure 73

Effect of Weight and Size On Truck Fuel Economy

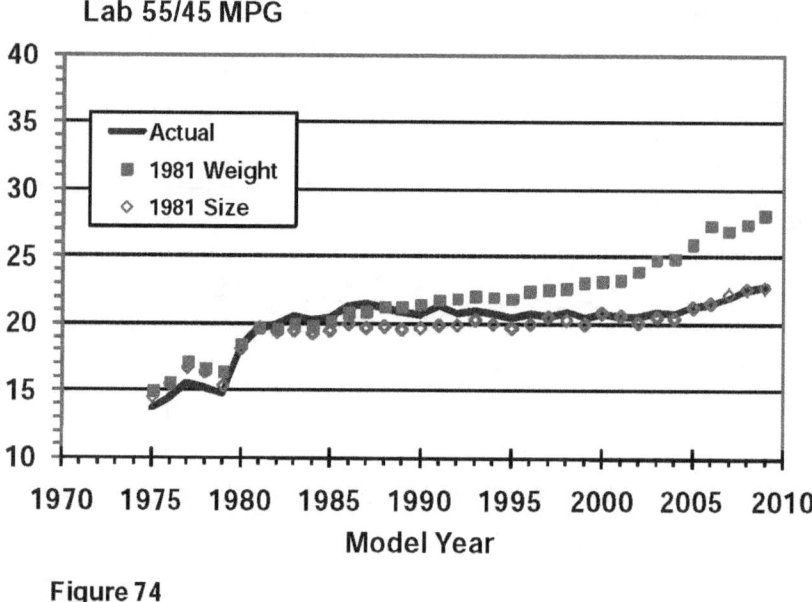

Figure 74

Effect of Weight and Size On Car Fuel Economy

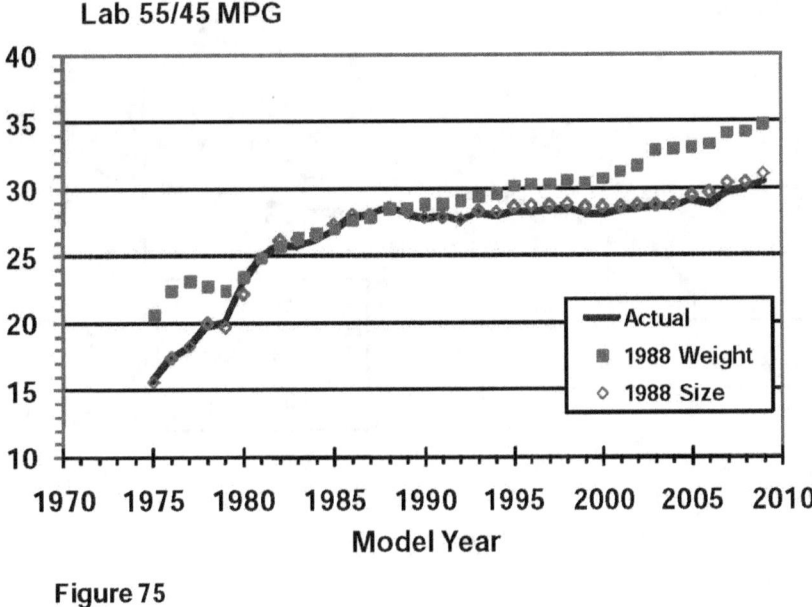

Figure 75

Effect of Weight and Size On Truck Fuel Economy

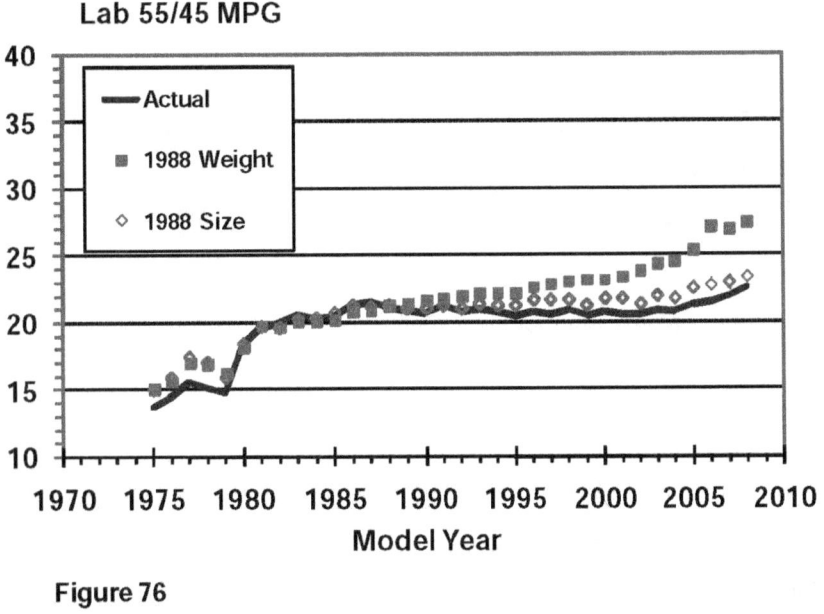

Figure 76

EPA-420-R-09-014   94   November 2009

IX. References

1. "U.S. Environmental Protection Agency, Fuel Economy and Emission Control," November 1972.

2. "Passenger Car Fuel Economy - Trends and Influencing Factors," SAE Paper 730790, Austin and Hellman, September 1973.

3. "Fuel Economy of the 1975 Models," SAE Paper 740970, Austin and Hellman, October 1974.

4. "Passenger Car Fuel Economy Trends Through 1976," SAE Paper 750957, Austin and Service, October 1975.

5. "Light-Duty Automotive Fuel Economy Trends Through 1977," SAE Paper 760795, Murrell, Pace, Service, and Yeager, October 1976.

6. "Light-Duty Automotive Fuel Economy Trends Through 1978," SAE Paper 780036, Murrell, February 1978.

7. "Light-Duty Automotive Fuel Economy Trends Through 1979," SAE Paper 790225, Murrell, February 1979.

8. "Light-Duty Automotive Fuel Economy Trends Through 1980," SAE Paper 800853, Murrell, Foster and Bristor, June 1980.

9. "Light-Duty Automotive Fuel Economy Trends Through 1981," SAE Paper 810386, Foster, Murrell and Loos, February 1981.

10. "Light-Duty Automotive Fuel Economy Trends Through 1982," SAE Paper 820300, Cheng, LeBaron, Murrell, and Loos, February 1982.

11. "Why Vehicles Don't Achieve EPA MPG On the Road and How That Shortfall Can Be Accounted For," SAE Paper 820791, Hellman and Murrell, June 1982.

12. "Light-Duty Automobile Fuel Economy Trends through 1983," SAE Paper 830544, Murrell, Loos, Heavenrich, and Cheng, February 1983.

13. "Passenger Car Fuel Economy - Trends Through 1984," SAE Paper 840499, Heavenrich, Murrell, Cheng, and Loos, February 1984.

14. "Light Truck Fuel Economy - Trends through 1984," SAE Paper 841405, Loos, Cheng, Murrell and Heavenrich, October 1984.

15. "Light-Duty Automotive Fuel Economy - Trends Through 1985," SAE Paper 850550, Heavenrich, Murrell, Cheng, and Loos, March 1985.

16. "Light-Duty Automotive Trends Through 1986," SAE Paper 860366, Heavenrich, Cheng, and Murrell, February 1986.

17. "Trends in Alternate Measures of Vehicle Fuel Economy," SAE Paper 861426, Hellman and Murrell, September 1986.

EPA-420-R-09-014

18. "Light-Duty Automotive Trends Through 1987," SAE Paper 871088, Heavenrich, Murrell, and Cheng, May 1987.

19. "Light-Duty Automotive Trends Through 1988," U.S. EPA, EPA/AA/CTAB/88-07, Heavenrich and Murrell, June 1988.

20. "Light-Duty Automotive and Technology Trends Through 1989," U.S. EPA, EPA/AA/CTAB/89-04, Heavenrich, Murrell, and Hellman, May 1989.

21. "Downward Trend in Passenger Car Fuel Economy--A View of Recent Data," U.S. EPA, EPA/AA/CTAB/90-01, Murrell and Heavenrich, January 1990.

22. "Options for Controlling the Global Warming Impact from Motor Vehicles," U.S. EPA, EPA/AA/CTAB/89-08, Heavenrich, Murrell, and Hellman, December 1989.

23. "Light-Duty Automotive Technology and Fuel Economy Trends through 1990," U.S. EPA, EPA/AA/CTAB/90-03, Heavenrich and Murrell, June 1990.

24. "Light-Duty Automotive Technology and Fuel Economy Trends through 1991," U.S. EPA/AA/CTAB/91-02, Heavenrich, Murrell, and Hellman, May 1991.

25. "Light-Duty Automotive Technology and Fuel Economy Trends through 1993," U.S. EPA/AA/TDG/93-01, Murrell, Hellman, and Heavenrich, May 1993.

26. "Light-Duty Automotive Technology and Fuel Economy Trends through 1996," U.S. EPA/AA/TDSG/96-01, Heavenrich and Hellman, July 1996.

27. "Light-Duty Automotive Technology and Fuel Economy Trends through 1999," U.S. EPA420-R-99-018, Heavenrich and Hellman, September 1999.

28. "Light-Duty Automotive Technology and Fuel Economy Trends 1975 through 2000," U.S. EPA420-R-00-008, Heavenrich and Hellman, December 2000.

29. "Light-Duty Automotive Technology and Fuel Economy Trends 1975 through 2001," U.S. EPA420-R-01-008, Heavenrich and Hellman, September 2001.

30. "Light-Duty Automotive Technology and Fuel Economy Trends 1975 through 2003," U.S. EPA420-R-03-006, Heavenrich and Hellman, April 2003.

31. "Light-Duty Automotive Technology and Fuel Economy Trends 1975 through 2004," U.S. EPA420-R-04-001, Heavenrich and Hellman, April 2004.

32. "Light-Duty Automotive Technology and Fuel Economy Trends 1975 through 2005," U.S. EPA420-R-05-001, Robert M. Heavenrich, July 2005.

33. "Light-Duty Automotive Technology and Fuel Economy Trends 1975 through 2006," U.S. EPA420-R-06-011, Robert M. Heavenrich, July 2006.

34. "Light-Duty Automotive Technology and Fuel Economy Trends: 1975 through 2007," U.S. EPA420-S-07-001, Office of Transportation and Air Quality, September 2007.

35. "Light-Duty Automotive Technology and Fuel Economy Trends: 1975 Through 2008," U.S. EPA420-R-08-015, Office of Transportation and Air Quality, September 2008.

36. "Concise Description of Auto Fuel Economy in Recent Years," SAE Paper 760045, Malliaris, Hsia and Gould, February 1976.

37. "Automotive Engine – A Future Perspective," SAE Paper 891666, Amann, 1989.

38. "Regression Analysis of Acceleration Performance of Light-Duty Vehicles," DOT HS 807 763, Young, September 1991.

39. "Determinates of Multiple Measures of Acceleration," SAE Paper 931805, Santini and Anderson, 1993.